I0043595

Edwin Lankester

The Advancement of Science

Occasional essays & addresses

Edwin Lankester

The Advancement of Science
Occasional essays & addresses

ISBN/EAN: 9783337034665

Printed in Europe, USA, Canada, Australia, Japan

Cover: Foto ©berggeist007 / pixelio.de

More available books at **www.hansebooks.com**

THE ADVANCEMENT OF SCIENCE

THE

ADVANCEMENT OF SCIENCE

OCCASIONAL ESSAYS & ADDRESSES

BY

E. RAY LANKESTER, M.A., LL.D., F.R.S.

London

MACMILLAN AND CO.

AND NEW YORK

1890

PREFACE

I HAVE to thank Mr. Knowles of the *Nineteenth Century*, Mr. Quilter of the *Universal Review*, and the Messrs. Black of the *Encyclopædia Britannica*, for permission to include in this volume articles originally published by them.

The collection which Messrs. Macmillan have kindly undertaken to republish deals with a variety of topics, and is, perhaps, wanting in unity of design. These essays and lectures have, however, all originated in my desire to promote the interests of science, either by explaining some of its later advances to the general public, or by showing the need for increased aid to scientific research from the State. Hence the somewhat comprehensive title which I have ventured to give to this little book. I hope that each paper in its turn may find some readers who will think that it possesses sufficient interest to justify its reappearance in its present form. Some of the papers now reprinted were published sufficiently long ago for changes to have occurred in the position of matters of which they treat.

In some cases a prediction has been verified, in others an appeal has been successful; and in regard to some few matters touched on, new views have taken the place of those adopted when my statement was originally made. I have added notes explanatory of such changes where they are actually necessary.

E. Ray Lankester.

December 1889.

CONTENTS

I

DEGENERATION:

A CHAPTER IN DARWINISM

DELIVERED AS ONE OF THE EVENING LECTURES AT THE SHEFFIELD
MEETING OF THE BRITISH ASSOCIATION FOR THE ADVANCEMENT OF
SCIENCE, 22D AUGUST 1879, UNDER THE PRESIDENCY OF PROFESSOR
ALLMAN, LL.D., F.R.S.

DEGENERATION:

A CHAPTER IN DARWINISM

IT is the misfortune of those who study that branch
of science which our President has done so much to
advance—I mean the science of living things—that
they are not able, in the midst of a vast assembly,
to render visible to all eyes the actual phenomena
to which their inquiries are directed. Whilst the
physicist and the chemist are able to make evident to
the senses of a great meeting the very things of which
they have to tell, the zoologist cannot hope ever to
share with those who form his audience the keen
pleasure of observing a new or beautiful organism;
he cannot demonstrate by means of actual specimens
the delicate arrangements of structure which it is his
business to record, and upon which he bases his conclusions. It is for this reason that he who would
bring to the notice of laymen some matter which at
the moment is occupying the attention of biological
students, must appear to be unduly devoted to speculation—hypothesis—to support which he cannot produce the facts themselves but merely the imperfect
substitutes afforded by pictures. It is perhaps not
altogether a matter for regret that there should be in

one great branch of science, as there is in biology, so
very marked a disproportion between the facilities for
demonstrating facts and the general interest attaching
to the theories connected with those facts.   We may
be thankful that at the present day we are not likely,
in the domain of biology, to make the mistake (which
has been made under other circumstances) of substituting the mere inspection and cataloguing of natural
objects for that more truly scientific attitude which
consists in assigning the facts which come under our
observation to their causes, or, in other words, to their
places in the order of nature.   Though we may rightly
object to the attempt which is sometimes made to
decry the modern teachings of biology as not being
" exact science," yet we may boldly admit the truth
of the assertion that we biologists are largely occupied
with speculations, hypotheses, and other products of
the imagination.   All true science deals with speculation and hypothesis, and acknowledges as its most
valued servant—its indispensable ally and helpmeet
—that which our German friends[1] call " Phantasie "
and we " the Imagination."   Our science, biology,
is not less exact—our conclusions are no less accurate
because they are only *probably* true. They are " probably true " with a degree of probability of which
we are fully aware, and which is only somewhat less
than the probability attaching to the conclusions
of other sciences which are commonly held to be
" exact."

These remarks are addressed to an Association for

[1] See Note A.

the advancement of science—of science which flourishes
and progresses by the aid of suppositions and the
working of the imagination. The Association has
been holding its annual sitting in various parts of
the British Islands for more than thirty years, and
yet it is still a very common and widely spread notion
that science, that is to say, *true* science according to
those who hold the notion, does not countenance
hypotheses, and sternly occupies itself with the exact
record of fact./ On the other hand, there are many
persons who run to an opposite extreme, and call by
the name of science any fanciful attempt to deal with
or account for a certain class of phenomena. The
words "science" and "scientific" are used so vaguely
and variously that one might almost come to the
conclusion that it would be well for our Association
to plainly state what is that thing for the advancement of which its meetings are held. I cannot venture
to speak in the name of my colleagues ; and no doubt
a review of the work done by the Association would
most fitly explain what that body understands by the
word "science." At the same time it is permissible
to take this opportunity of briefly stating what science
is and what it is not, so far as I am able to judge of
the fitting use of the word.

Science is certainly *not* any and every kind of
knowledge. Knowledge of literature, of the beautiful
things which have been written or otherwise produced
by human ingenuity, is not science. Knowledge of
the various manufacturing processes in use by civilised
men is not science ; nor knowledge of the names of

the stars, or of the joints of a beetle's leg. Science
cannot be identified with knowledge of any particular
class of objects, however detailed that knowledge may
be. It is a common mistake to consider all knowledge
of raw products, of living objects or other natural
objects, as necessarily "science." The truth is, that
a man may have great knowledge of these things as
so many facts, and yet be devoid of "science." And,
on the other hand, that which is properly called
science embraces not only such subject - matter as
that just alluded to, but also may find its scope in
the study of language, of human history, and of the
workings of the human mind.

The most frequent and objectionable misuse of the
word "science" is that which consists in confounding
science with invention—in applying the term which
should be reserved for a particular kind of knowledge
to the practical applications of that knowledge. Such
things as electric lighting and telegraphs, the steam-
engine, gas, and the smoky chimneys of factories, are
by a certain school of public teachers, foremost among
whom is the late Oxford Professor of Fine Art, persistently ascribed to science, and gravely pointed out
as the pestilential products of a scientific spirit. They
are, in fact, nothing of the kind. American inventors
and electric lamps, together with all the factories in
Sheffield, might be obliterated without causing a
moment's concern to a single student of science. It
is of the utmost importance for the progress and well-
being of science that this should be understood ; that
the eager, practical spirit of the inventor who gains

large pecuniary rewards by the sale of his inventions
should not be confounded with what is totally different
and remote from it, namely, the devoted, searching
spirit of science, which, heedless of pecuniary rewards,
ever faces nature with a single purpose—*to ascertain
the causes of things.* It seems to me impossible to
emphasise too strongly in such a place and in such a
meeting as this, that Invention is widely separate
from, though dependent on Science. Invention is
worldly-wise, and despises the pursuit of knowledge
for its own sake. She awaits the discoveries of
Science, in order to sell them to civilisation, gathering the golden fruit which she has neither planted
nor tended. Invention follows, it is true, the foot-
steps of Science, but at a distance: she is utterly
devoid of that thriftless yearning after knowledge,
that passionate desire to know the truth, which causes
the unceasing advance of her guide and benefactress.

We may, it seems to me, say that of all kinds and
varieties of knowledge that only is entitled to the
name "science" which can be described as Knowledge
of Causes, or Knowledge of the Order of Nature. It
is this knowledge to which the great founder of
European science—Aristotle the Greek—pointed as
true knowledge: τότε ἐπιστάμεθα ὅταν τὴν αἰτίαν εἴδωμεν.
Science is that knowledge which enables us to demonstrate so far as our limited faculties permit, that the
appearances which we recognise in the world around
us are dependent in definite ways on certain properties
of matter: science is that knowledge which enables,
or tends to enable us, to assign its true place in the

series of events constituting the universe, to any and
every thing which we can perceive.

The method by which scientific knowledge is
gained—knowledge of the causes of nature—is precisely the same as that by which knowledge of causes
in everyday life is gained. Something—an appearance—has to be accounted for: the question in both
cases is, "Through what cause, in relation to what
antecedent is this appearance brought about?" In
scientific inquiry, as in everyday life, a hypothesis—
a provisional answer or guess—is the reply ; and the
truth of that guess or hypothesis is then tested. This
testing is an essential part of the process. "If my
guess be true, then so-and-so as to which I can decide
by inspection or experiment, must be true also," is
the form which the argument takes, and the inquiry is
thus brought to a point where observation can decide
the truth of the hypothesis or first guess. In everyday life we have often to be content without fully
testing the truth of our guesses, and hurry into action
based on such unverified suppositions. Science, on
the other hand, can always wait, and demands again
and again the testing and verification of guesses
before they are admitted as established truths fit to
be used in the testing of new guesses and the building
up of scientific doctrine.

The delicately-reared imaginations of great investigators of natural things have from time to time given
birth to hypotheses—guesses at truth—which have
suddenly transformed a whole department of knowledge, and made the causes of things quite clear

which before seemed likely to remain always concealed.
So great is the value of hypothesis, so essential to
scientific discovery, that the most skilled and highly
trained observer may spend his life in examining and
scrutinising natural objects and yet fail, if he is not
guided by hypothesis, to observe particular facts which
are of the uttermost importance for the explanation of
the causes of the things which he is studying. Nature,
it has been said, gives no reply to a general inquiry—
she must be interrogated by questions which already
contain the answer she is to give ; in other words, the
observer can only observe that which he is led by hypothesis to look for : the experimenter can only obtain
the result which his experiment is designed to obtain.

For a long time the knowledge of living things, of
plants, and of animals could hardly be said to form
part of the general body of science, for the causes of
these things were quite unknown. They were kept
apart as a separate region of nature, and were supposed to have been pitched, as it were, into the midst
of an orderly and cause-abiding world without cause
or order : they were strangers to the universal harmony prevailing around them. Fact upon fact was
observed and recorded by students of plants and
animals, but having no hypothesis as to the causes of
what they were studying, the naturalists of twenty
years ago, and before that day, though they collected
facts, made slow progress and some strange blunders.
Suddenly one of those great guesses which occasionally appear in the history of science, was given to the
science of biology by the imaginative insight of that

greatest of living naturalists—I would say that greatest of living men—Charles Darwin.

In the form in which Mr. Darwin presented his view to the world it was no longer a mere guess. He had already tried it and proved it in an immense series of observations ; it had already been converted by twenty years' labour on his part into an established doctrine, and the twenty years which have passed since he published the Origin of Species have only served to confirm, by thousands of additional tests, the truth of his original guess.

Space will not allow me to go fully into the history of the Darwinian theory, but it is necessary for my present purpose that I should state precisely what that theory is. It involves a number of subordinate hypotheses which, together with the main hypothesis, furnish us with a complete "explanation," as it is called, of the facts which have been ascertained as to living things ; in other words, it assigns living things to their causes, gives them their place in the Order of Nature.

It is a very general popular belief at the present day that the Darwinian theory is simply no more than a capricious and anti-theological assertion that mankind are the modified descendants of ape-like ancestors.

Though most of my readers, I do not doubt, know how imperfect and erroneous a conception this is, yet I shall not, I think, be wasting time in stating what the Darwinian theory really is. In fact, it is so continuously misrepresented and misunderstood, that no opportunity should be lost of calling attention to its real character. Bit by bit, naturalists had succeeded in discovering the order of nature—so far that all the great facts of the universe, the constitution and movements of the heavenly bodies, the form of our earth, and all the peculiarities of its crust, had been successfully assigned to one set of causes—the properties of matter, which are set forth in what we know by the name of the "laws of physics and chemistry." Whilst geologists, led by Lyell, had shown that the strata of the earth's crust and its mountains, rivers, and seas were due to the long-continued operation of the very same general causes —the physico-chemical causes—which at this moment are in operation and are continuing their work of change, yet the living matter on the crust of the earth had to be excluded from the grand uniformity which was elsewhere complete.

The first hypothesis, then, which was present to Mr. Darwin's mind, as it had been to that of other earlier naturalists, was this: "Have not all the varieties or species of living things (man, of course, included) been produced by the continuous operation of the same set of physico-chemical causes which alone we can discover, and which alone have been proved sufficient to produce everything else?" "If this be so," Mr. Darwin must have argued (and here it was that he boldly stepped beyond the speculations of Lamarck and adopted the method by which Lyell had triumphantly established Geology as a science), "these causes must still be able to produce

new forms, and are doing so wherever they have
opportunity." He had accordingly to bring the
matter to the test of observation by seeking for
some case of the production of new forms of plants,
or of animals, by natural causes at the present day.
Such cases he found in the production of new forms
or varieties of plants and animals, by breeders.
Breeders (the persons who make it their business
to produce new varieties of flowers, of pigeons, of
sheep, or what not) make use of two fundamental
properties of living things in order to accomplish
their purpose. These two properties are, firstly, that
no two animals or plants, even when born of the
same parents, are exactly alike; this is known as
Variation: secondly, that an organism, as a rule,
inherits, that is to say, is born with the peculiarities
of its parents; this is known as Transmission, and
is simply dependent on the fact that the offspring
of any plant or animal is only a detached portion
of the parent—a chip of the old block, as the saying
is. The breeder selects from a number of specimens
of a plant or animal a variety which comes nearest to
the form he wishes to produce. Supposing he wished
to produce a race of oxen with short horns, he would
select from his herd bulls and cows with the shortest
horns, and allow these only to breed; they would
transmit their relatively short horns to their offspring,
and from these again the cattle with the shortest
horns would be selected by the breeder for propagation,
and so on through several generations. In the end a
very short-horned generation would be obtained, differ-

ing greatly in appearance from the cattle with which the breeder started.

Now we know of no facts which forbid us to suppose that could a breeder continue his operations indefinitely for any length of time—say for a few million years—he could convert the short-horned breed into a hornless breed; that he could go on and thicken the tail, could shorten the legs, get rid of the hind limbs altogether by a series of insensible gradations, and convert the race into forms like the Sirenia, or sea-cows. But if he could do this, you have only to give him a longer time still and there is no obstacle remaining to the conversion, by the same kind of process, of a polyp into a worm, or of a worm into a fish, or even of a monkey into a man.[1]

So far we have supposed the interference of a breeder who selects and determines the varieties which shall propagate themselves; so far we have not got a complete explanation, for we must find a substitute in nature for the human selection exercised by the breeder. The question arises, then, "Is there any necessary selective process in nature which could have operated through untold ages, and so have represented the selective action of the breeder, during an immense period of time?" Strangely enough, Mr. Darwin was led to the discovery of such a cause existing necessarily in the mechanical arrangements of nature, by reading the celebrated book of an English clergyman, the Rev. Mr. Malthus, On Population. On happening to read this book, Mr. Darwin himself tells us that the idea

[1] See Note B.

of "natural selection" flashed upon him. That idea is as follows. Not only among mankind, but far more largely among other kinds of animals and among plants, the number of offspring produced by every pair is immensely in excess of the available amount of the food appropriate to the particular species in question. Accordingly, there is necessarily a struggle for existence—a struggle among all those born for the possession of the small quantum of food. The result of this struggle is to pick out, or select, a few who survive and propagate the species, whilst the majority perish before reaching maturity. The fact that no two members of a species are alike has already been shown to be the starting-point which enables the breeder to make *his* selection. So, too, with *natural* selection in the struggle for existence ; the fact that all the young born of one species are not exactly alike—but some larger, some smaller, some lighter, some darker, some short-legged, some big-eyed, some long-tongued, some sharp-toothed, and so on—furnishes the opportunity for a selection. Those varieties which are best fitted to obtain food and to baffle their competitors, gain the food and survive, the rest perish.

We have, then, to note that the hypothesis that there must be a selection—which was framed or deduced as a "test hypothesis" from the earlier hypothesis that species have arisen by the action of causes still competent to produce new forms—led Mr. Darwin to the discovery of this great cause—the "natural selection," or "survival of the fittest," in the struggle for existence. Just as the breeder can slowly change the proportions of the animals or plants on which he operates, so in inconceivably long periods of time has this struggling of varieties, and the consequent natural selection of the fittest, led to the production, from shapeless primitive living matter, of all the endless varieties of complicated plants and animals which now people the world. Countless varieties have died out, leaving only their modified descendants to puzzle the ingenuity of the biologist.

Of the tens and hundreds of thousands of intermediate forms we know nothing by direct observation. They have perished as better-fitted forms ousted them in the never-ending conflict. But we feel sure that they once were in existence, and can infer what was their structure, and what were their peculiarities, by the study of the structure and attributes of their now living descendants.

If all the forms of life at present living are the modified offspring of a smaller number of ancestral forms which have died out, and if these again were the modified descendants produced by ordinary parentage of a single original living thing—then the whole series of forms that have ever lived could, if we had them before us, be arranged in the form of a great family-tree—the various branches presenting a perfect gradation of forms arranged one after another, leading down from the terminal twigs (which would represent the latest forms produced) to larger and larger branches, until the common trunk representing the original ancestor would be reached. Our actual means of observing the genealogical affinities of different kinds of animals and plants may be understood by a further use of the metaphor of a genealogical tree in shape like an elm or an oak. Suppose the genealogical tree completely written out—a perfect record—to be sunk in muddy water so that only its topmost branches and twigs are here and there visible—then you have a fair notion of the present condition of the great family of organisms. Only the topmost twigs remain visible, the rest of the great family-tree of living beings is hidden from view, submerged beneath the muddy waters of time. Naturalists have, however, undertaken to reconstruct this great genealogical tree. It is a main object now in the study both of zoology and of botany to find out what are the cousinships, what the exact genetic relationships of all the various species of plants and animals, and so to show, even to the minutest detail, in what particular ways physico-chemical causes have brought about and modified the forms of living things.

The task is not quite so difficult as the comparison to a submerged forest-tree would lead one to expect; at the same time it is more difficult than those who have boldly attempted it appear to believe. We have one great help in the carefully worked-out systematic classification of animals and plants according to their structure. We are justified in assuming as a general law that animals or plants of like structure have descended from common ancestors—that is to say, that the same kind of organisation (especially where a number of elaborate details of structure are involved) has not been twice produced by natural selection. Thus

we are entitled to conclude that all the animals which
have a backbone and pharyngeal gill-slits combined—
the Vertebrates, as we call them—have descended
from a common parent; that all the animals with a
muscular foot-like belly and lateral gill filaments, the
Molluscs, have also had a common parent, and so on.

A classification according to structure goes then a
long way towards mapping out the main lines of the
family-tree of organisms. We are further assisted in
the task by the fossil remains of extinct organisms
which sometimes give to us the actual ancestors of
forms now living. But the most remarkable aid to
the correct building up of the pedigree of animals at
least (and the remarks which follow are confined to
that division of the organic world) is afforded by the
changes—the phases of development—which every
animal exhibits in passing from the small shapeless
egg to the adult condition. The aid which we here
obtain depends on the following facts. Just as we
suppose any one animal—say a dog—to have developed
by slow change through an immense series of ancestors
which become simpler and simpler as we recede into
the past until we reach a small shapeless lump of living
matter devoid of structure, so do we find actually as
a matter of fact, which any one can see for himself,
that every individual animal begins its individual life
as a structureless particle which is thrown off from its
parent, and is known as the egg-cell (Fig. 1). Gradually passing through a series of more and more elaborated conditions of structure, that egg grows into

the adult dog. The changes which have taken countless ages in the one case, are accomplished in a few
weeks in the other.

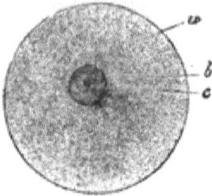

FIG. 1.—An egg: a single
corpuscle of protoplasm
with nucleus *b c*, and
body *a*.

And now we have to note the
important fact which makes this
process of development so intensely
interesting in relation to the pedigree of the animal kingdom. There
is very strong reason to believe
that it is a general law of trans-
mission or inheritance, that structural characteristics
appear in the growth of a young organism in the order
in which those characteristics have been acquired by
its ancestors. At first the egg of a dog represents
(imperfectly, it is true) in form and structure the
earliest ancestors of the dog ; a few days later it has
the form and structure of somewhat later ancestors ;
later still the embryo dog resembles less remote
ancestors ; until at last it reaches the degree of
elaboration proper to its immediate forefathers.

Accordingly the phases of development or growth
of the young are a brief recapitulation of the phases of
form through which the ancestors of the young creature
have passed. In some animals this recapitulation is
more, in others it is less complete. Sometimes the
changes are hurried through and disguised, but we
find here and there in these histories of growth from
the egg most valuable assistance in the attempt to
reconstruct the genealogical tree. The history of the
development of the common frog is a good illustration
of the *kind* of evidence in question.

The frog's egg first gives rise to a little aquatic creature with external gills and a tail—the tadpole—which gradually loses its gills and its tail and acquires in their place lungs and four legs (Fig. 2), so as now

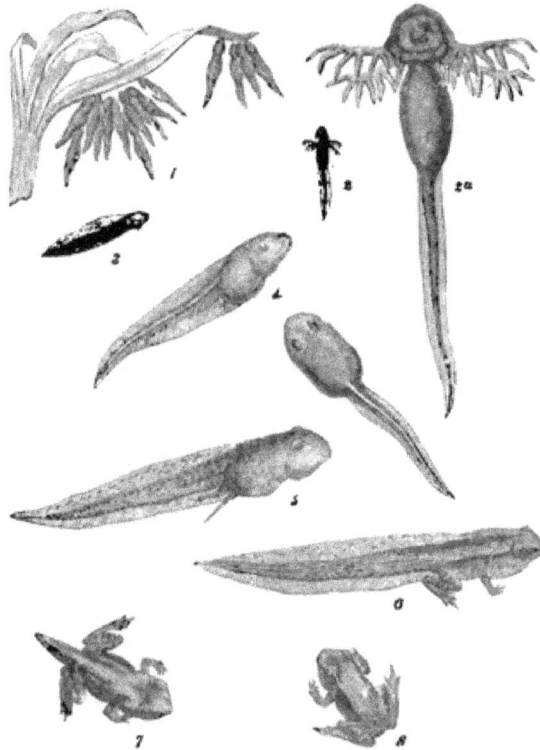

FIG. 2.—Tadpoles and young of the Common Frog. 1, Recently hatched (twice natural size) ; 2 and 2a, same enlarged to show the external gills ; 3 and 4, later stages with gill-slits covered by a membrane leaving only the spiracle (see Fig. 16) as an exit for the respired water ; 5, with hind legs appearing ; 6, with both fore and hind legs ; 7, atrophy of the tail ; 8, young frog.

to be fitted for life on dry land. From what we otherwise know of the structure of the frog and the animals to which it is allied, we are justified in concluding that the tadpole is a recapitulative phase of development,

and represents to us more or less closely an ancestor
of the frog which was provided with gills and tail in
the adult state, and possessed neither legs nor lungs.

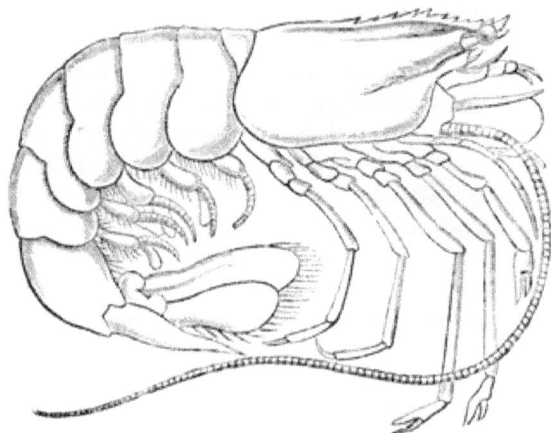

FIG. 3.—Adult shrimp of the genus Peneus.

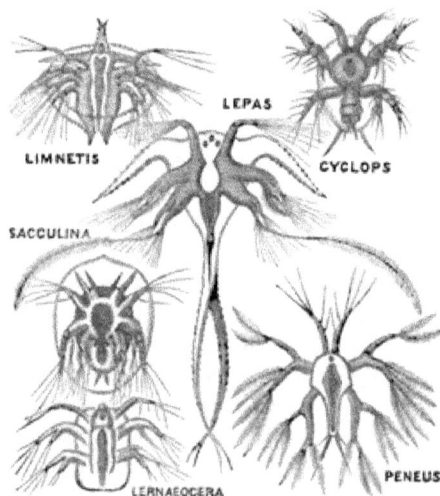

FIG. 4.—Nauplius larval-form of various Crustacea (Shrimps, Water-fleas,
Barnacles, etc.)

A less familiar case is that of a certain kind of
shrimp, which is illustrated in the woodcuts (Fig. 3
and right lower corner of Fig. 4). The little creature

which issues from the egg of this shrimp is known as
the "Nauplius form." Many animals very different
in appearance from this shrimp make their first appearance in the world as Nauplii; and it appears probable that the Nauplius phase is the recapitulative representation of an ancestor common to all this set of

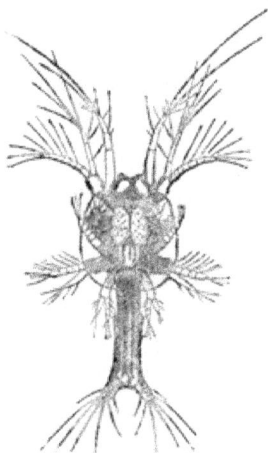

FIG. 5.—Larva of the Shrimp
Peneus.

FIG. 6.—More advanced larva
of the Shrimp Peneus.

animals, an ancestor which was not exactly like the
Nauplius, but not very different from it.

The Nauplius of our shrimp gradually elongates.
At first it has but three pair of limbs, but it soon
acquires additional pairs, and a jointed body, and thus
by gradually adding to its complexity of structure as
seen in Figs. 5 and 6, it approximates more and more
to the adult form from the egg of which it originated.

And now we are approaching the main point to

which I wish to draw the reader's attention. In attempting to reconstruct the pedigree of the animal kingdom and so to exhibit correctly the genetic relationships of all existing forms of animals, naturalists have hitherto assumed that the process of natural selection and survival of the fittest has invariably acted so as either to improve and elaborate the structure of all the organisms subject to it, or else has left them unchanged, exactly fitted to their conditions, maintained as it were in a state of balance. It has been held that there have been some six or seven great lines of descent—main branches of the pedigree—such as that of the Vertebrates, that of the Molluscs, that of the Insects, that of the Starfish, and so on ; and that along each of these lines there has been always and continuously a progress—a change in the direction of greater elaboration.

Each of these great branches of the family-tree is held to be independent—they all branch off nearly simultaneously from the main trunk like the leading branches of an oak. The animal forms constituting the series in each of these branches are supposed to gradually increase in elaboration of structure as we pass upwards from the main trunk of origin and climb farther and farther towards the youngest, most recent twigs. New organs have, it is supposed, been gradually developed in each series, giving their possessors greater powers, enabling them to cope more successfully with others in that struggle for existence in virtue of which these new organs have been little by little called into being. At the

same time here and there along the line of march,
certain forms have been supposed to have "fallen
out," to have ceased to improve, and being happily
fitted to the conditions of life in which they were
long ago existing, have continued down to the
present day to exist in the same low, imperfect
condition. It is in this way that the lowest forms
of animal life at present existing are usually explained, such as the microscopic animalcules, Amœbæ
and Infusoria. It is in this way that the lower or
more simply-made families of higher groups have
been generally regarded. The simpler living Mollusca
or shellfish have been supposed necessarily to represent the original forms of the great race of Mollusca.
The simpler Vertebrates have been supposed necessarily to represent the original Vertebrates. The
simpler Worms have been supposed necessarily to
be the stereotyped representatives of very ancient
Worms.

That this is, to a certain extent, a true explanation of the existence at the present day of low forms
of animals is proved by the fact that we find in very
ancient strata fossil remains of animals which differ,
ever so little, from particular animals existing at the
present day; for instance, the Brachiopods (lamp-
shells), Lingula and Terebratula, the King-crabs,
and the Pearly Nautilus are found living at the
present day, and are also found with no appreciable
difference in very ancient strata of the earth's crust;
strata deposited so long ago that most of the forms
of life at present inhabiting the earth's surface had

not then been brought into existence, whilst other
most strange and varied forms occupied their place,
and have now for long ages been extinct.

Whilst we are thus justified by the direct testi-
mony of fossil remains in accounting for *some* living
forms on the hypothesis that their peculiar conditions
of life have been such as to maintain them for an
immense period of time *in statu quo* unchanged, we
have no reason for applying this hypothesis, and this
only, to the explanation of all the more imperfectly
organised forms of animal or plant life.

It is clearly enough possible for a set of forces
such as we sum up under the head "natural selection" to so act on the structure of an organism as
to produce one of three results, namely, these : to
keep it *in statu quo ;* to increase the complexity of
its structure ; or lastly, to diminish the complexity
of its structure. We have as possibilities either
BALANCE, or ELABORATION, or DEGENERATION.

Owing, as it seems, to the predisposing influence
of the systems of classification in ascending series
proceeding steadily upwards from the "lower" or
simplest forms to the "higher" or more complex
forms,—systems which were prevalent before the
doctrine of transformism had taken firm root in the
minds of naturalists, there has been up to the present
day an endeavour to explain every existing form of
life on the hypothesis that it has been maintained
for long ages in a state of Balance ; or else on
the hypothesis that it has been Elaborated, and is
an advance, an improvement, upon its ancestors.

Only one naturalist—Dr. Dohrn, of Naples—has put forward the hypothesis of Degeneration as capable of wide application to the explanation of existing forms of life ;[1] and his arguments in favour of a general application of this hypothesis have not, I think, met with the consideration which they merit.

The statement that the hypothesis of Degeneration has not been recognised by naturalists generally as an explanation of animal forms, requires to be corrected by the exception of certain kinds of animals, namely, those that are parasitic or quasi-parasitic. With regard to parasites, naturalists have long recognised what is called retrogressive metamorphosis ; and parasitic animals are as a rule admitted to be instances of Degeneration. It is the more remarkable whilst the possibility of a degeneration—a loss of organisation making the descendant far *simpler* or *lower* in structure than its ancestor—has been admitted for a few exceptional animals, that the same hypothesis should not have been applied to the explanation of other simple forms of animals. The hypothesis of Degeneration will, I believe, be found to render most valuable service in pointing out the true relationships of animals which are a puzzle and a mystery when we use only and exclusively the hypothesis of Balance, or the hypothesis of Elaboration. It will, as a true scientific hypothesis, help us to discover causes.

We may now examine a few examples of un-

---

[1] " Der Ursprung der Wirbelthiere und das Princip des Functions-wechsels." Leipzig, 1875.

deniably degenerate animals, and first, I may call to mind the very remarkable series of lizard-like animals which exist in the south of Europe and in other countries, which exhibit in closely related genera a gradual loss of the limbs—a local or limited Degeneration. We have the common Lizard (*Lacerta*), with five toes on each of its well-grown fore and hind limbs ; then we have side by side with this a lizard-like creature, *Seps*, in which both pairs of limbs have become ridiculously small, and are evidently ceasing to be useful in the way in which those of *Lacerta* are useful ; and lastly, we have *Bipes*, in which the anterior pair of limbs has altogether vanished, and only a pair of stumps, representing the hinder limbs, remain.

No naturalist doubts that *Seps* and *Bipes* represent two stages of Degeneration, or atrophy of the limbs ; that they have, in fact, been derived from the five-toed four-legged form, and have lost the locomotor organs once possessed by their ancestors. This very partial or local atrophy is not, however, that to which I refer when using the word Degeneration. Let us imagine this atrophy to extend to a variety of important organs, so that not only the legs, but the organs of sense, the nervous system, and even the mouth and digestive organs are obliterated—then we shall have pictured a thorough-going instance of Degeneration.

Degeneration may be defined as a gradual change of the structure in which the organism becomes adapted to less varied and less complex conditions

of life ; whilst Elaboration is a gradual change of
structure in which the organism becomes adapted
to more and more varied and complex conditions of
existence.   In Elaboration there is a new expression
of form, corresponding to new perfection of work in
the animal machine.   In Degeneration there is suppression of form, corresponding to the cessation of
work.   Elaboration of some one organ *may* be a
necessary accompaniment of Degeneration in all the
others ; in fact, this is very generally the case ; and
it is only when the total result of the Elaboration of
some organs, and the Degeneration of others, is such
as to leave the whole animal in a lower condition,
that is, fitted to less complex action and reaction in
regard to its surroundings, than was the ancestral
form with which we are comparing it (either actually
or in imagination) that we speak of that animal as
an instance of Degeneration.

Any new set of conditions occurring to an animal
which render its food and safety very easily attained,
seem to lead as a rule to Degeneration ; just as an
active healthy man sometimes degenerates when he
becomes suddenly possessed of a fortune ; or as Rome
degenerated when possessed of the riches of the ancient
world.   The habit of parasitism clearly acts upon
animal organisation in this way.   Let the parasitic
life once be secured, and away go legs, jaws, eyes, and
ears ; the active, highly-gifted crab, insect, or annelid
may become a mere sac, absorbing nourishment and
laying eggs.

Reference was made above to the larval stage of

a certain shrimp (Figs. 4, 5, 6). Let us now compare
these with the young stages of a number of shrimp-
like animals, viz. Sacculina, Lernæocera, Lepas, Cy-
clops, Limnetis (all drawn in Fig. 4), some of which
lead a parasitic life. All start equally with the recapitulative phase known as the Nauplius ; but whilst

Fig. 7.—Adult Sacculina
(female).

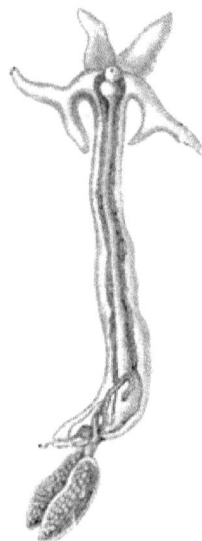

Fig. 8.—Adult female
Lernæocera.

the Nauplius of the free-living shrimp grows more and
more elaborate, observe what happens to the parasites
—they degenerate into comparatively simple bodies ;
and this is true of their internal structure as well as
of their external appearance. The most utterly reduced of these parasites is the curious Sacculina
(Fig. 7) which infests Hermit-crabs, and is a mere
sac filled with eggs, and absorbing nourishment from
the juices of its host by root-like processes.

Lernæocera again, which in the adult condition is found attached to the gills of fishes, has lost the well-developed legs of its Nauplius childhood and become an elongated worm-like creature (Fig. 8), fitted only to suck in nourishment and carry eggs.

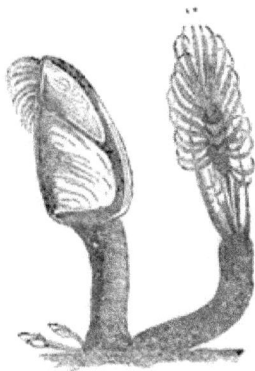

FIG. 9.—Adult Barnacle or Lepas (one of the Cirrhipedes). Natural size.

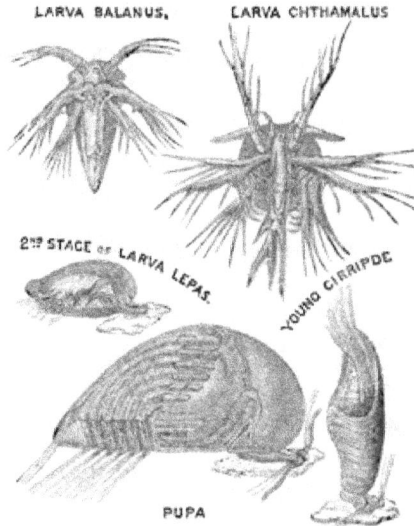

FIG. 10.—Development of Cirrhipedes (Barnacle and Sea-acorn). After Huxley.

Amongst these Nauplii—all belonging to the great group Crustacea, which includes crabs and shrimps—is one which gives rise to an animal decidedly degenerate, but not precisely parasitic in its habits. This Nauplius is the young of the ship's Barnacle, a curious stalked body, enclosed in a shell of many pieces (Fig. 9). The egg of the Barnacle gives rise to an actively swimming Nauplius, the history of which is very astonishing. After swimming about for a time the Barnacle's Nauplius fixes its head against a piece of wood, and takes to a perfectly fixed, immobile state of life (Fig. 10). The upper figures represent the Nauplius stage of animals closely resembling the Barnacle; the lower figures show the transformation of the Nauplius into the young Barnacle. Its organs of touch and of sight atrophy, its legs lose their locomotor function, and are simply used for bringing floating particles to the orifice of the stomach; so that an eminent naturalist has compared one of these animals to a man standing on his head and kicking his food into his mouth.

Were it not for the recapitulative phases in the development of the Barnacle, we may doubt whether naturalists would *ever* have guessed that it was a degenerate Crustacean. It was, in fact, for a long time regarded as quite remote from them, and placed among the snails and oysters; its true nature was only admitted when the young form was discovered.

Other parasitic organisms, which exhibit extreme degeneration as compared with their free-living relatives, might be cited and figured in profusion, did our limits permit. Very noteworthy are the degenerate Spiders—the mites, leading to still more degenerate forms, the Linguatulæ.[1]

We have two of these represented in Figs. 11 and 12. The one (Fig. 11), as compared with a spider, is

[1] I have, since the above was written, applied the principle of degeneration to the explanation of the living representatives of the group Arachnida, and shewn that the King Crab (Limulus) is the nearest representative of the ancestors of scorpions, spiders, and mites, and in fact must be classed with them. Professor Claus of Vienna has, some years later, adopted similar views. December 1889.

seen still to possess the eight walking legs, small, it is true, whilst the palps and daggers of the spider have dwindled to a beak projecting from the front of the globular unjointed body.   In the other the eight legs have become mere stumps, and the body is elongated like that of a worm.

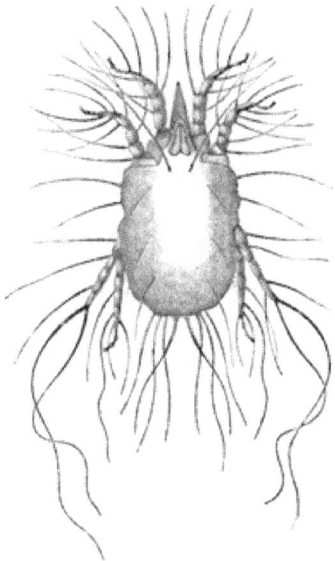

FIG. 11.   FIG. 12.

FIG. 11.—Acarus equi.   A degenerate Spider or mite, parasitic on the skin of the horse.

FIG. 12.—Degenerate Spider (Demodex folliculorum) found in the skin of the human face.

The instances of degeneration which we have so far examined are due to parasitism, except in the example of the Barnacle, where we have an instance of degeneration due to sessile and immobile habit of life. We may now proceed to look at some sessile or immobile animals which are not usually regarded as degenerate, but which, I think, there is every reason to

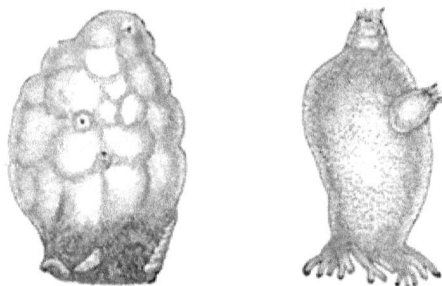

FIG. 13.—Two adult Ascidians : to the left Phallusia—to the right Cynthia : the incurrent and excurrent orifices are seen as two prominences. Half the natural size.

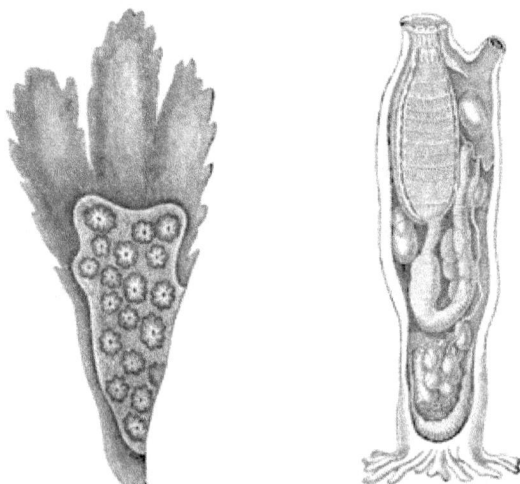

FIG. 14.   FIG. 15.

FIG. 14.—A colony of compound Ascidians (Botryllus) growing on a piece of sea-weed (Fucus). Each star corresponds to eight or more conjoined As-cidians. Natural size.

FIG. 15.—Anatomy of an Ascidian (Phallusia). At the top is the mouth, to the right the orifice of the cloaca. In the cloaca lies an egg, and above it the oblong nerve-ganglion. The perforated pharynx follows the mouth and leads to the bent intestine which is seen to open into the cloaca. The space around the curved intestine is the body-cavity ; in it are seen oval bodies, the eggs, and quite at the lower end the curved heart. The root-like processes at the base serve to fix the Ascidian to stones, shells, or weed.

believe are the degenerate descendants of very much

higher and more elaborate ancestors. These are certain marine animals, the Ascidians, or sea-squirts. These animals are found encrusting rocks, stones, and weeds on the sea-bottom. Sometimes they are solitary (Fig. 13), but many of them produce buds, like plants, and so form compound masses or sheets of individuals all connected and continuous with one another, like the buds on a creeping plant (Fig. 14).

We will examine one of the simple forms—a tough mass like a leather bottle with two openings; water is continually passing in at the one and out at the other of these apertures. If we remove the leathery outer case (Fig. 15), we find that there is a soft creature within which has the following parts: Leading from the mouth a great throat, followed by an intestine. The throat is perforated by innumerable slits, through which the water passes into a chamber—the cloaca: in passing, the water aërates the blood which circulates in the framework of the slits. The intestine takes a sharp bend, which causes it to open also into the cloaca. Between the orifice of the mouth and of the cloaca there is a nerve-ganglion.

My object in the next place is to show that the structure and life-history of these Ascidians may be best explained on the hypothesis that they are instances of degeneration; that they are the modified descendants of animals of higher, that is, more elaborate structure, and in fact are degenerate Vertebrata, standing in the same relation to fishes, frogs, and men, as do the barnacles to shrimps, crabs, and lobsters.

The young of some, but by no means of all these
Ascidians, have a form totally different from that of
their parents.   The egg of Phallusia gives rise to
a tadpole, a drawing of which placed side by side with
the somewhat larger tadpole of the common frog is
seen in the adjoining figure (Fig. 16).   The young
Ascidian has the same general shape as the young frog,
but not only this ;   the resemblance extends into
details, the internal organs agreeing closely in the two
cases.   Further still, as shown by the beautiful researches of the Russian naturalist, Kowalewsky, the

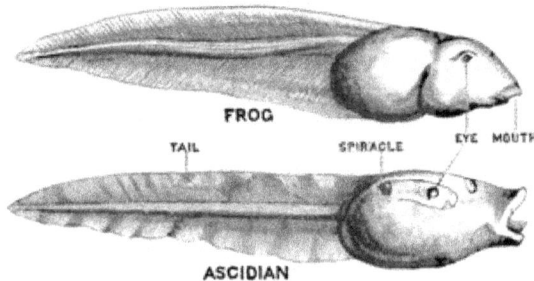

FIG. 16.—Tadpole of Frog and of Ascidian.   Surface view.

resemblance reaches absolute identity when we examine
the way in which the various organs arise from the
primitive egg-cell.   Tail, body, spiracle, eye, and
mouth agree in the two tadpoles, the only important
difference being in the position of the two mouths and
in the fact that the Ascidian has one eye while the frog
has two.

Now let us look at the internal organs (Fig. 17).
There are four structures, which are all four possessed
at some time of their lives by all those animals which
we call the Vertebrata, the great branch of the pedi-

gree to which fishes, reptiles, birds, beasts, and men
belong. And the combination of these marks or
structural peculiarities is an overwhelming piece of
evidence in favour of the supposition that the creatures
which possess this combination are derived from one
common ancestor. Just as one would conclude that a
man whom one might meet, say on Salisbury Plain,
must belong to the New Zealand race, if it were found
not only that he had the colour, and the hair, and the
shape of head of a New Zealander, but also that he

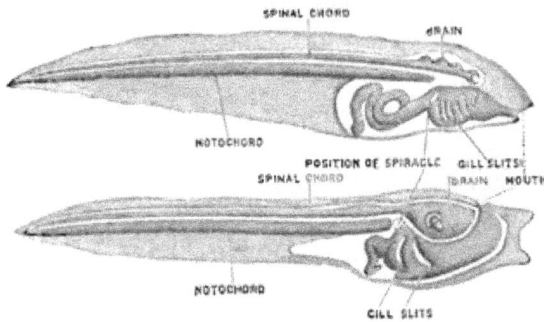

Fig. 17.—Tadpole of Frog and of Ascidian. Diagrams representing the chief
internal organs.

was tattooed like a New Zealander, carried the weapons
of a New Zealander, and, over and above, in addition
to these proofs, that he talked the Maori language and
none other ; so here, in the case of the vertebrate race,
there are certain qualities and possessions, the accumulation of which cannot be conceived of as occurring
in any animal but one belonging to that race. These
four great structural features are—first, the primitive
backbone or notochord ; second, the throat perforated
by gill-slits ; third, the tubular nerve-centre or spinal
cord and brain placed along the back ; and lastly, and

perhaps most distinctive and clinching as an evidence
of affinity, the myelonic or cerebral eye.

Now let us convince ourselves that these four
features exist not only in the frog's tadpole, as they do
in all fishes, reptiles, birds, and beasts, but that they
also exist in the Ascidian tadpole, and, it may be
added, coexist in no other animals at all.

The corresponding parts are named in Figs. 16 and
17, in such a way as to render their agreement tolerably
clear, whilst in Fig. 18 a more detailed representation
of the head of an Ascidian tadpole is given.

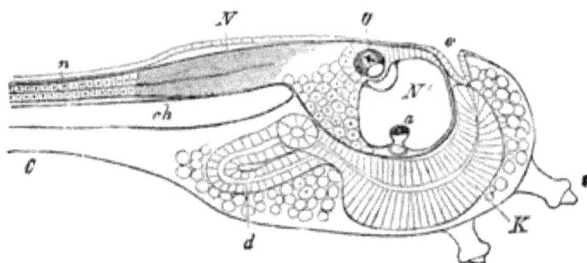

FIG. 18.—Ascidian Tadpole with a part only of the tail C.   N, nervous system
     with the enlarged brain in front and the narrow spinal chord behind (n); N',
     is placed in the cavity of the brain ; O, the single cerebral eye lying in the
     brain ; a, similarly placed auditory organ ; K, pharynx ; d, intestine ; o,
     rudiment of the mouth ; ch, notochord or primitive backbone. (From
     Gegenbaur's *Elements of Comparative Anatomy*.)

It is clear then that the Ascidians must be admitted
to be Vertebrates, and must be classified in that great
sub-kingdom or branch of the animal pedigree.   The
Ascidian tadpole is very unlike its parent the Ascidian,
and has to go through a process of *degeneration* in
order to arrive at the adult structure.   The diagrams
which are reproduced in Figs. 19 and 20 show how
this degeneration proceeds.   It will be observed, that

in somewhat the same manner as the young barnacle, the young Ascidian fixes itself to a stone by its head;

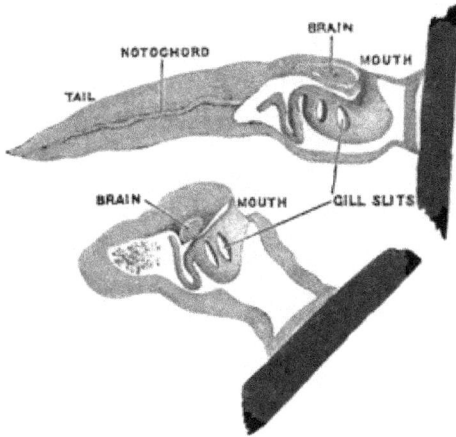

Fig. 19.—Degeneration of Ascidian Tadpole to form the adult.   The black pieces represent the rock or stone to which the Tadpole has fixed its head.

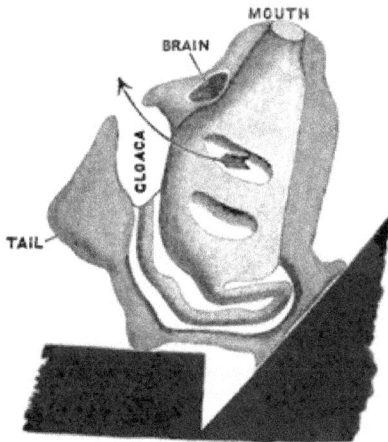

Fig. 20.—Very young Ascidian with only two gill-slits.   Compare with Fig. 15 ; which is, however, seen from the other side, so that left there corresponds to right here.

then the tail with its notochord and nerve-chord atrophies.   The body grows and gradually changes its

shape, whilst the cloacal chamber forms.   The brain
remains quite small and undeveloped, and the remarkable myelonic eye (the eye in the brain) disappears.
The number of gill-slits increases as the animal grows
in size and its outer skin becomes tough and leather-
like.

Before saying anything further on the subject of
degeneration, it seems desirable once more to direct

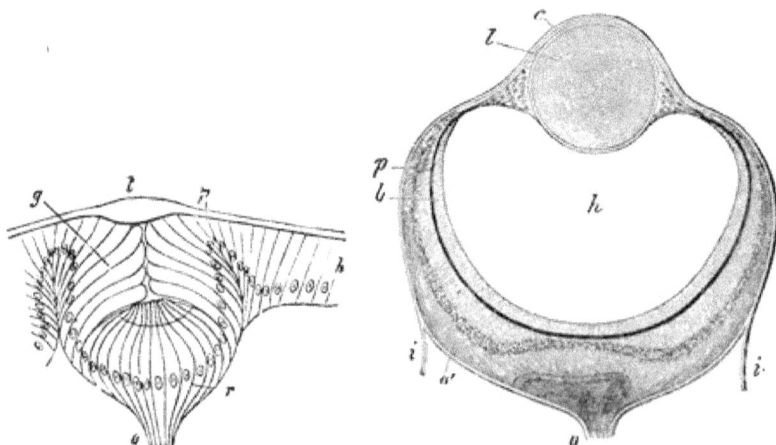

Fig. 21.   Fig. 22.

Fig. 21.—Section through the eye ("surface eye") of a Water-beetle's larva.
All the cells are seen to be in a row continuous with *h*, the cells of the outer-
most skin or ectoderm. *p*, pigmented cells ; *r*, retinal cells connected at *o*
with the optic nerve ; *g*, transparent cells (forming a kind of "vitreous
body") ; *l*, cuticular lens. (From Gegenbaur's *Elements of Comparative
Anatomy*, after Grenacher.)
Fig. 22.—Section through the eye ("surface-eye") of a Marine Worm (Neo-
phanta). *i*, integument spreading over the front of the eye *c* ; *l*, cuticular
lens ; *h*, cavity occupied by vitreous body ; *p*, retinal cells ; *b*, pigment ; *o*,
optic nerve ; *o'*, expansion of optic nerve.

attention to the myelonic or cerebral eye which the
Ascidian tadpole possesses in common with all Vertebrates.   All other animals which have eyes develop
the retina or sensitive part of the eye from their outer
skin (see Figs. 21 and 22, and explanation).   It is

easy to understand that an organ which is to be affected by the light should form on the surface of the body where the light falls.   It has long been known as a very puzzling and unaccountable peculiarity of Vertebrates, that the retina or sensitive part of the eye grows out in the embryo as a bud or vesicle of the brain, and thus forms deeply below the surface and away from the light (see Fig. 23, and explanation). The Ascidian tadpole helps us to understand this, for

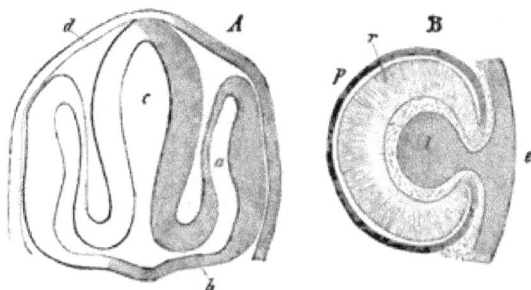

FIG. 23.—A. Vertical section through the head of a very young fish, showing in the centre the cavity of the brain c.   On each side is a hollow outgrowth (a) which will form the retina of the fish's eye ("cerebral eye "); b will become the optic nerve connecting the brain and the retina ; d, integument.—B. Later condition of the hollow outgrowth (a) of A.   Its outer wall r is pressed against its deeper wall p by an ingrowth (l) from the outer skin (ectoderm) e ; r gives rise to the retinal cells, whilst only l, the cellular lens, is derived from the surface of the skin.

it is perfectly transparent and has its eye actually *inside* its brain.   The light passes through the transparent tissues and acts on the pigmented eye, lying deep in the brain.   We are thus led to the suggestion —and I believe this inference to be now for the first time put into so many words—that the original Vertebrate must have been a transparent animal, and had an eye or pair of eyes *inside* its brain, like that of the

Ascidian tadpole. As the tissues of this ancestral Vertebrate grew denser and more opaque, the eye-bearing part of the brain was forced by natural selection to grow outwards towards the surface, in order that it might still be in a position to receive the influence of the sun's rays. Thus the very peculiar mode of development of the Vertebrate eye from two parts, a brain-vesicle (Fig. 23, *A a*, and *B p r*) and a skin-vesicle (Fig. 23, *B e, l*), is accounted for.[1]

The cases of degeneration which I have up to this point brought forward, are cases which admit of very little dispute or doubt. They are attested by either the history of the individual development of the organisms in question, as in Sacculina, in the Barnacle, and in the Ascidian, or they are cases where the comparison of the degenerate animal with others like it in structure, but not degenerate, renders the hypothesis of degeneration an unassailable one. Such cases are the Acarus or mite, and the skin-worm (Demodex).

We have seen that degeneration, or the simplification of the general structure of an animal, may be due to the ancestors of that animal having taken to one of two new habits of life, either the parasitic or the immobile. Other new habits of life appear also to be such as to lead to degeneration. Let us suppose a race of animals fitted and accustomed to catch their food, and having a variety of organs to help them in this chase—suppose such animals suddenly to acquire the

---

[1] I do not at the present time (November 1889) attach great importance to the above suggestion. The facts admit of other possible and plausible interpretations.

power of feeding on the carbonic acid dissolved in the water around them just as green plants do. This would lead to a degeneration; they would cease to hunt their food, and would bask in the sunlight, taking food in by the whole surface, as plants do by their leaves. Certain small flat worms, by name Convoluta, of a bright green colour, appear to be in this condition. Their green colour is known to be the same substance as leaf-green; and Mr. Patrick Geddes has recently shown that by the aid of this green substance they feed on carbonic acid, making starch from it as plants do. As a consequence we find that their stomachs and intestines as well as their locomotive organs become simplified, since they are but little wanted. These vegetating animals, as Mr. Geddes calls them, are the exact complement of the carnivorous plants, and show how a degeneration of animal forms may be caused by vegetative nutrition.

Another possible cause of degeneration appears to be the indirect one of minute size. It cannot be doubted that natural selection has frequently acted on a race of animals so as to reduce the size of the individuals. The smallness of size has been favourable to their survival in the struggle for existence, and in some cases they have been reduced to even microscopic proportions. But this reduction of size has, when carried to an extreme, resulted in the loss or suppression of some of the most important organs of the body. The needs of a very minute creature are limited as compared with those of a large one, and thus we may find heart and blood-vessels, gills and kidneys, besides legs and muscles, lost by the diminutive degenerate descendants of a larger race. That this is a possible course of change all will, I think, admit. It is actually exemplified in Appendicularia—the only adult representative of the Ascidian tadpole—still tadpole-like in form and structure, but curiously degenerate and simplified in its internal organs. This kind of degeneration is also exemplified in the Rotifers, or wheel animalcules, in the minute Crustacean water-fleas (Ostracoda), and in the Moss-polyps, or Polyzoa. Roughly then we may sum up the immediate antecedents of degenerative evolution as, 1, Parasitism ; 2, Fixity or immobility ; 3, Vegetative nutrition ; 4, Excessive reduction of size. This is not a logical enumeration, for each of these causes involves, or may be inseparably connected with, one or more of the others. It will serve for the present as well as a more exhaustive analysis. (See Note C.)

And now we have to note an important fact with regard to the evidence which we can obtain of the occurrence of this process of degeneration. We have seen that the most conclusive evidence is that of the recapitulative development of the individual. The Ascidian Phallusia shows itself to be a degenerate Vertebrate by beginning life as a tadpole. But such recapitulative development is by no means the rule. Quite arbitrarily, we find, it is exhibited in one animal and not in a nearly allied kind. Thus very many animals belonging to the Ascidian group have no tadpole young—just as some tree-frogs have no tadpoles. It is quite possible, and often, more often than not,

occurs that *the most important part* of the recapitulative phases are absent from the developmental history
of an animal.   The egg proceeds very rapidly to produce the adult form, and all the wonderful series of
changes showing the animal's ancestry are absolutely
and completely omitted ; that is to say, all those stages
which are of importance for our present purpose.   Just
as certain bodies pass from the solid to the liquid state
at a bound, omitting all intermediate phases of con-
sistence, but giving evidence of "internal work" by
the suggestive phenomenon of latent heat—so do these
embryos skip long tracts in the historically continuous
phases of form, and present to us only the intangible
correlative "internal work" in place of the tangible
series of embryonic changes of shape.

Now I want to put this case—a supposition—before
the reader who has so far followed me in these pages.
Suppose, as might well have happened, that the Barnacles, one and all, instead of recapitulating in their
early life, were to develop directly from the egg to the
adult form, as so many animals do ; should we have
ever made out that they were degenerate Crustaceans ?
Possibly we should : their adult structure still bears
important marks of affinities with crabs and shrimps ;
but as a matter of fact before their recapitulative development had been discovered they were classed by
the great Cuvier and other naturalists with the Molluscs,
the mussels and snails.

Suppose again that all the existing Ascidians, as
many of them actually have, had long ago lost their
recapitulative history in growth from the egg : suppose

that no such a thing as an Ascidian tadpole existed, but that the Ascidian's egg grew as directly as possible into an Ascidian, in every living species of the group. This might easily be the case. Then most assuredly we should not have the least notion that the Ascidians were degenerate Vertebrates. We should still class them where they used to be classed before the Russian naturalist Kowalewsky discovered the true history and structure of the Ascidian tadpole. I believe that I shall have the assent of every naturalist when I say that the vertebrate character of the Ascidians and the history of their degeneration would never have been suspected, or even dreamed of, had the Ascidian tadpoles ceased to appear in the course of the Ascidian development at a geological period anterior to the present epoch.

This being the case, it must be admitted that it is quite possible—I do not say more than possible—that other groups of animals besides parasites, Barnacles, and Ascidians, are degenerate. It is quite possible that animals with considerable complexity of structure, at least as complex as the Ascidians, may have been produced by degeneration from still more highly organised ancestors. Any group of animals to which we can turn may possibly be the result of degeneration, and yet offer no evidence of that degeneration in its growth from the egg.

Accordingly, wherever we can note that a group of organisms is characterised by habits likely to lead to degeneration, such as I have enumerated, viz. parasitism or immobility, or certain special modes of nutrition, or again, by minute size of its representatives —there we are justified in applying the hypothesis of degeneration, even in the absence of any confirmatory evidence from embryology. When we so apply this hypothesis we find in not a few cases, in working over the details of the organisation of many different animals by the light which it affords—that much becomes clear and assignable to cause which, on the hypothesis either of "balance" or of "elaboration," is quite hopelessly obscure. As examples of groups of animals which can thus be satisfactorily explained I may cite first of all the Sponges : as only somewhat less degenerate, we have all the Polyps and Coral-animals, also the Star-fishes. Amongst the Mollusca—the group of head-less bivalves, the oysters, mussels and clams, known as the Lamellibranchs, are, when one once looks at their structure in this light, clearly enough explained as degenerated from a higher type of head-bearing active creatures like the Cuttle-fish; whilst the Polyzoa or Moss-polyps stand in precisely the same kind of relation to the higher Mollusca [1] as do the Ascidians to the higher Vertebrates: they have greatly degenerated, and become minute encrusting organisms which, like some of the Ascidians, build up colonies by plant-like budding growth. The Rotifers, or wheel animalcules, I have already mentioned as best explained by the supposition that they are the descendants of far larger and more fully-organised animals provided with loco-

---

[1] Not to the higher Mollusca in all probability, but to some higher worm-like form. The ancestry of the Polyzoa is still a profound mystery! December, 1889.

motive appendages or limbs : they have dwindled and
degenerated to their present minute size and curiously
suggestive structure.

Besides these there are other very numerous cases
of animal structure which can best be explained by
the hypothesis of degeneration. A discussion of these,
and a due exposition of the application of the hypothesis of degeneration to the various groups just cited,
would involve a complete treatise on comparative
anatomy and embryology, and lead far beyond the
limitations of this little volume.

All that has been, thus far, here said on the subject
of Degeneration is so much zoological specialism,
and may appear but a narrow restriction of the discussion to those who are not zoologists. Though we
may establish the hypothesis most satisfactorily by
the study of animal organisation and development, it
is abundantly clear that degenerative evolution is by
no means limited in its application to the field of
zoology. It clearly offers an explanation of many
vegetable phenomena, and is already admitted by
botanists as the explanation of the curious facts connected with the reproductive process in the higher
plants. As a further example of its application in
this field, the yeast-plant may be adduced, which is
in all probability a degenerate floating form derived
from a species of mould. In other fields, wherever
in fact the great principle of evolution has been
recognised, degeneration plays an important part.
In tracing the development of languages, philo-
logists have long made use of the hypothesis of

degeneration. Under certain conditions, in the
mouths and minds of this or that branch of a race,
a highly elaborate language has sometimes degenerated and become no longer fit to express complex
or subtle conceptions, but only such as are simpler
and more obvious. (See Note D.)

The traditional history of mankind furnishes us
with notable examples of degeneration. High states
of civilisation have decayed and given place to low
and degenerate states. At one time it was a favourite
doctrine that the savage races of mankind were degenerate descendants of the higher and civilised races.
This general and sweeping application of the doctrine
of degeneration has been proved to be erroneous by
careful study of the habits, arts, and beliefs of savages;
at the same time there is no doubt that many savage
races as we at present see them are actually degenerate and are descended from ancestors possessed
of a relatively elaborate civilisation. As such we
may cite some of the Indians of Central America,
the modern Egyptians, and even the heirs of the
great oriental monarchies of præ-Christian times.
Whilst the hypothesis of universal degeneration as
an explanation of savage races has been justly discarded, it yet appears that degeneration has a very
large share in the explanation of the condition of
the most barbarous races, such as the Fuegians, the
Bushmen, and even the Australians. They exhibit
evidence of being descended from ancestors more
cultivated than themselves.

With regard to ourselves, the white races of

Europe, the possibility of degeneration seems to be worth some consideration. In accordance with a tacit assumption of universal progress—an unreasoning optimism—we are accustomed to regard ourselves as necessarily progressing, as necessarily having arrived at a higher and more elaborated condition than that which our ancestors reached, and as destined to progress still further. On the other hand, it is well to remember that we are subject to the general laws of evolution, and are as likely to degenerate as to progress. As compared with the immediate forefathers of our civilisation—the ancient Greeks—we do not appear to have improved so far as our bodily structure is concerned, nor assuredly so far as some of our mental capacities are concerned. Our powers of perceiving and expressing beauty of form have certainly not increased since the days of the Parthenon and Aphrodite of Melos. In matters of the reason, in the development of intellect, we may seriously inquire how the case stands. Does the reason of the average man of civilised Europe stand out clearly as an evidence of progress when compared with that of the men of bygone ages? Are all the inventions and figments of human superstition and folly, the self-inflicted torturing of mind, the reiterated substitution of wrong for right, and of falsehood for truth, which disfigure our modern civilisation—are these evidences of progress? In such respects we have at least reason to fear that we may be degenerate. Possibly we are all drifting, tending to the condition of intellectual Barnacles or Ascidians. It is possible

for us—just as the Ascidian throws away its tail and
its eye and sinks into a quiescent state of inferiority
—to reject the good gift of reason with which every
child is born, and to degenerate into a contented life
of material enjoyment accompanied by ignorance
and superstition. The unprejudiced, all-questioning
spirit of childhood may not inaptly be compared to
the tadpole tail and eye of the young Ascidian ; we
have to fear lest the prejudices, preoccupations,
and dogmatism of modern civilisation should in any
way lead to the atrophy and loss of the valuable
mental qualities inherited by our young forms from
primæval man.

There is only one means of estimating our position, only one means of so shaping our conduct that
we may with certainty avoid degeneration and keep
an onward course. We are as a race more fortunate
than our ruined cousins—the degenerate Ascidians.
For us it is possible to ascertain what will conduce
to our higher development, what will favour our
degeneration. To us has been given the power " to
know the causes of things," and by the use of this
power it is possible for us to control our destinies.
It is for us by ceaseless and ever hopeful labour
to try to gain a knowledge of man's place in the
order of nature. When we have gained this fully
and minutely, we shall be able by the light of the
past to guide ourselves in the future. In proportion as the whole of the past evolution of civilised
man, of which we at present perceive the outlines, is
assigned to its causes, we and our successors on the

globe may expect to be able duly to estimate that which makes for, and that which makes against, the progress of the race. The full and earnest cultivation of Science—the Knowledge of Causes—is that to which we have to look for the protection of our race —even of this English branch of it—from relapse and degeneration.

NOTES

A

"Die Phantasie ist ein unentbehrliches Gut; denn sie ist es, durch welche neue Combinationen zur Veranlassung wichtiger Entdeckungen gemacht werden. Die Kraft der Unterscheidung des isolirenden Verstandes, sowohl als die der erweiternden und zum Allgemeinen strebenden Phantasie sind dem Naturforscher in einem harmonischen Wechselwirken nothwendig. Durch Störung dieses Gleichgewichts wird der Naturforscher von der Phantasie zu Träumereien hingerissen, während diese Gabe den talentvollen Naturforscher von hinreichender Verstandesstärke zu den wichtigsten Entdeckungen führt."—Johannes Müller, *Archiv für Anatomie*, 1834.

B

To many persons the conclusion that man is the naturally modified descendant of ape-like ancestors appears to be destructive of the belief in an immortal soul, and in the teachings of Christianity; and accordingly they either reject Darwinism altogether, or claim for man a special exemption from the mode of origin admitted for other animals.

It seems worth while, in order to secure a calm
and unprejudiced consideration for the teachings of
Darwinism, to point out to such persons that, as a
matter of fact, whatever views we may hold with
regard to a soul and the Christian doctrines, they
cannot be in the smallest degree affected by the admission that man has been derived from ape-like
ancestors by a process of natural selection, so long as
the demonstrable fact, not denied by any sane person,
is admitted, namely, that every individual man grows
by a process of natural modification from a homogeneous egg-cell or corpuscle.   Assuredly it cannot lower
our conception of man's dignity if we have to regard
him as "the flower of all the ages," bursting from the
great stream of life which has flowed on through
countless epochs with one increasing purpose, rather
than as an isolated, miraculous being, put together
abnormally from elemental clay, and cut off by such
portentous origin from his fellow-animals, and from
that gracious Nature to whom he yearns with filial
instinct, knowing her, in spite of fables, to be his dear
mother.

A certain number of thoughtful persons admit the
development of man's body by natural processes from
ape-like ancestry, but believe in the non-natural
intervention of a Creator at a certain definite stage in
that development, in order to introduce into the
animal which was at that moment a man-like ape,
something termed "a conscious soul," in virtue of
which he became an ape-like man.   It appears to me
perfectly legitimate and harmless for individuals to

make such an assumption if their particular form of philosophy or of religion requires it. Such an assumption does not in any way traverse the inferences from facts to which Darwinism leads us; at the same time zoological science does not, and cannot be expected to, give any support to such an assumption. The gratuitous and harmless nature of the assumption so far as zoological science is concerned, and accordingly the baselessness of the hostility to Darwinism of those who choose to make it, may be seen by the consideration of a parallel series of facts and assumptions, which puts the matter clearly enough in its true light.

No one ventures to deny, at the present day, that every human being grows from the egg *in utero*, just as a dog or a monkey does; the facts are before us and can be scrutinised in detail. We may ask of those who refuse to admit the gradual and natural development of man's consciousness in the ancestral series, passing from ape-like forms into indubitable man, "How do you propose to divide the series presented by every individual man in his growth from the egg? At what particular phase in the embryonic series is the soul with its potential consciousness implanted? Is it in the egg? in the fœtus of this month or of that? in the new-born infant? or at five years of age?" This, it is notorious, is a point upon which Churches have never been able to agree; and it is equally notorious that the unbroken series exists —that the egg becomes the fœtus, the fœtus the child, and the child the man. On the other hand we have the historical series—the series, the existence of

which is inferred by Darwin and his adherents. This
is a series leading from simple egg-like organisms to
ape-like creatures, and from these to man. Will
those who cannot answer our previous inquiries
undertake to assert dogmatically in the present case
at what point in the historical series there is a break
or division? At what step are we to be asked to
suppose that the order of nature was stopped, and a
non-natural soul introduced? The philosopher or
theologian of this or that school may arbitrarily draw
an imaginary line here or there in either series, and
the evolutionist will not raise a finger to stop him.
As long as truth in the statement of fact, and logic
in the inference from observed fact are respected,
there need be no hostility between evolutionist and
theologian. The theologian is content in the case of
individual development from the egg to admit the
facts of individual evolution, and to make assumptions
which lie altogether outside the region of scientific
inquiry. So, too, it would seem only reasonable that
he should deal with the historical series, and frankly
accept the natural evolution of man from lower
animals, declaring dogmatically, if he so please, but
not as an inference of the same order as are the
inferences of science, that something called the soul
arrived at any point in the series which he may think
suitable. At the same time, it would appear to be
sufficient, even for the purposes of the theologian, to
hold that whatever the two above-mentioned series
of living things contain or imply, they do so as the
result of a natural and uniform process of development,

that there has been one "miracle" once and for all time. It should not be a ground of offence to any school of thinkers, that Darwinism, whilst leaving them free scope, cannot be made actually contributory to the support of their particular tenets.

The difficulties which the theologian has to meet when he is called upon to give some account of the origin and nature of the soul, certainly cannot be said to have been increased by the establishment of the Darwinian theory. For from the earliest days of the Church, ingenious speculation has been lavished on the subject. As to the origin of the individual soul, Tertullian tells us as follows, *De Anima*, chap. xix. : "Anima velut surculus quidam ex matrice Adami in propaginem deducta, et genitalibus semine foveis commodata. Pullulabit tam intellectu quam et sensu."

Whilst St. Augustine says : "Harum autem sententiarum quatuor de anima, utrum de propagine veniant, an in singulis quibusque nascentibus mox fiant, an in corpora nascentium jam alicubi existentes vel mittantur divinitus, vel sua sponte labantur, *nullam temere affirmari oporteret :* aut enim nondum ista quæstio a divinorum librorum catholicis tractatoribus, pro merito suæ obscuritatis et perplexitatis, evoluta atque illustrata est ; aut si jam factum est, nondum in manus nostras hujuscemodi litteræ provenerunt."

C

A VERY important form of degeneration, not touched on in the text, is that exhibited in the Mexican axolotl, where the larval form of a Salamander develops generative organs, and is arrested in its further progress to the adult parental form.   It is not possible to class this with the other phenomena which I have enumerated as Degeneration, since there is no modification of an adult structure, but simple arrest, and retention of the larval structure in all its completeness. I should call the phenomenon exhibited by the axolotl "arrest" or "super-larvation" rather than degeneration.

The result of super-larvation is in so far similar to that of those changes to which it is desirable to restrict the term "degeneration," that it may be classed under "simplificative evolution" as opposed to "elaborative evolution."   That there is a very real difference between super-larvation and degeneration may best be seen by taking a case of each process and instituting a comparison.   The axolotl proceeds regularly on its course of development from the egg, but instead of passing from the aquatic gilled condition to the terrestrial gill-less adult form of the Salamander, it remains arrested in the earlier condition, develops its reproductive organs, and propagates itself.   There is no loss or atrophy in this case, but simply a dead stop in a progressive course.   On the other hand, as we have seen, the Ascidian loses, by a process of atrophy and destruction, a powerful locomotive organ, a highly-developed eye, a relatively large nervous system.   The former may be compared to a permanent childishness, the latter to the second childhood, which is really atrophy and decay.   It is highly probable that super-larvation has taken place at various epochs and in various groups of the animal kingdom, just as it does in the axolotl, and yet we cannot hope for evidence fitted to establish its occurrence in any one case, where it is no longer possible by exceptional conditions to recover (as in the case of the axolotl, which can experimentally be made to advance to the Salamander phase by proper treatment), the discarded, more developed adult form. By super-larvation it would be possible for an embryonic form developed in relation to special embryonic conditions and not re-capitulative of an ancestry, to become the adult form of the race, and thus to give to the subsequent evolu-tion of that race a totally different and otherwise improbable direction.

It seems also exceedingly probable that super-larva-tion does not occur only as in the axolotl through premature maturation of the reproductive organs, but the phenomenon *may* develop itself more slowly by a gradual creeping forward, as it were, of larval features. Just as the adaptations acquired in, and having rela-tion to, later life tend to show themselves in an early period of the development of the individual and out of due season ; so do characters acquired by the early embryo, and having relation only to this early period of life tend to remain as permanent structures, and by their invasion to perturb the adult organisation. Such perturbation may tend either to simplification or elaboration.

D

THE term (degeneration of language) includes two very distinct things; the one is degeneration of grammatical form, the other degeneration of the language as an instrument of thought. The former is a far commoner phenomenon than the latter, and, in fact, whilst actually degenerating so far as grammatical complexity is concerned, a language may be at the same time becoming more and more serviceable, or more and more perfect as an organ having a particular function. The decay of useless inflections and the consequent simplification of language may be compared to the specialisation of the one toe of the primitively five-toed foot of the horse, whilst the four others which existed in archaic horses are, one by one, atrophied. Taken by itself, this phenomenon may possibly be described as degeneration, but inasmuch as the whole horse is not degenerate, but, on the contrary, specialised and elaborated, it is advisable to widely distinguish such local atrophy from general degeneration. In the same way language cannot, in relation to this question, be treated as a thing by itself—it must be regarded as a possession of the human organism, and the simplification of its structure merely means in most cases its more complete adaptation to the requirements of the organism.

True degeneration of language is therefore only found as part and parcel of a more general degeneration of mental activity. To some extent the conclu-

sion that this or that language, as compared with its
earlier condition, exhibits evidence of such degeneration, must be matter of taste and open to discussion.
For instance, the English of Johnson may be regarded
as degenerate when compared with that of Shake-
speare. There is less probability of a difference of
opinion as to the degeneracy of modern Greek as
compared with " classical " Greek; or of some of the
modern languages of Hindustan as compared with
Sanskrit, and I am informed that the same kind of
degeneration is exhibited by modern Irish as compared
with old Irish. Degeneration, in the proper sense of
the word, so far as it applies to language, would seem
to mean simply a decay or diversion of literary taste
and of literary production in the race to which such
language may be appropriate.

II

BIOLOGY AND THE STATE

The President's Address to the Biological Section of the British Association at Southport, 1883.

BIOLOGY AND THE STATE

IT has become the custom for the presidents of the various Sections of this Association to open the proceedings of the departments with the chairmanship of which they are charged, by formal addresses. In reflecting on the topics which it might be desirable for me to bring under your notice, as your president, on the present occasion, it has occurred to me that I might use this opportunity most fitly by departing somewhat from the prevailing custom of reviewing the progress of science in some special direction during the past year, and that, instead of placing before you a summary of the results recently obtained by the investigations of biologists in this or that line of inquiry, I might ask your attention and that of the external public (who are wont to give some kindly consideration to the opinions expressed on these occasions) to a matter which is even more directly connected with the avowed object of our Association, namely, "the Advancement of Science." I propose to place before you a few observations upon the provision which exists in this country for the advancement of that branch of science to which Section D is dedicated—namely, Biology.

I am aware that it is usual for those who speak of men of science and their pursuits to ignore altogether such sordid topics as the one which I have chosen to bring forward. A certain pride on the one hand, and a willing acquiescence on the other hand, usually prevent those who are professionally concerned with scientific pursuits from exposing to the public the pecuniary destitution and the consequent crippling and languor of scientific research in this country. Those Englishmen who take an interest in the progress of science are apt to suppose that, in some way which they have never clearly understood, the pursuit of scientific truth is not only its own reward, but also a sufficient source of food, drink, and clothing. Whilst they are interested and amused by the remarkable discoveries of scientific men, they are astonished whenever a proposal is mentioned to assign salaries to a few such persons sufficient to enable them to live decently while devoting their time and strength to investigation. The public are becoming more and more anxious to have the opinion or report of scientific men upon matters of commercial importance, or in relation to the public health; and yet in ninety-nine cases out of a hundred they expect to have that opinion for the asking, although accustomed to pay other professional men handsomely for similar service. There is, it appears, in the public mind a vague belief that men who occupy their time with the endeavour to add to knowledge in this or that branch of science are mysteriously supported by the State Exchequer, and are thus fair game for attacking with all sorts of demands for gratuitous service ; or, on the other hand, the notion at work appears sometimes to be that the making of new knowledge—in fact, scientific discovery —is an agreeable pastime, in which some ingenious gentlemen, whose business in other directions takes up their best hours, find relaxation after dinner or on the spare hours of Sunday. Such mistaken views ought to be dispelled with all possible celerity and determination. It is in part owing to the fact that the real state of the case is not widely and persistently made known to the public, that no attempt is made in this country to raise scientific research, and especially biological research, from the condition of destitution and neglect under which it suffers—a condition which is far below that of these same interests in France and Germany, and even in Holland, Belgium, Italy, and Russia, and is discreditable to England in proportion as she is richer than other States.

It appears to me that, in placing this matter before you, I may remove myself from any suggestion of self-interest by at once stating that the great defect to which I shall draw your attention is not that the few existing public positions which are open in this country to men who intend to devote their chief energies to biological research are endowed with in-sufficient salaries, but that there is not anything like a sufficiently large number of those posts, and that there is in that respect, from a national point of view, a pecuniary starvation of biology, a withholding of money which (to use another metaphor) is no less the sinews of the war of science against ignorance than of other less glorious campaigns. Surely men engaged in the scientific profession may advocate the claim of science to maintenance and needful pecuniary provision! It seems to me that we should, if necessary, stifle, rather than be controlled by, that pride which tempts us to paint the scientific career as one far above and independent of pecuniary considerations; whereas all the while we know that knowledge is languishing, that able men are drawn off from scientific research into other careers, that important discoveries are approached and their final grasp relinquished, that great men depart and leave no disciples or successors, simply for want of that which is largely given in other countries, of that which is most abundant in this country, and is so lavishly expended on armies and navies, on the development of commercial resources, on a hundred injurious or meaningless charities—viz. money.

I have no doubt that I have the sympathy of all my hearers in wishing for more extensive provision in this country for the prosecution of scientific research, and especially of biological research. I need hardly remind this audience of the almost romantic history of some of the great discoveries which have been made in reference to the nature and history of living things, during the past century. The microscope, which was a drawing-room toy a hundred years ago, has, in the hands of devoted and gifted students of nature, been the means of giving us knowledge which, on the one

hand, has saved thousands of surgical patients from
terrible pain and death, and, on the other hand, has
laid the foundation of that new philosophy with which
the name of Darwin will for ever be associated. When
Ehrenberg and, later, Dujardin described and figured
the various forms of Monas, Vibrio, Spirillum, and
Bacterium which their microscopes revealed to them,
no one could predict that fifty years later these
organisms would be recognised as the cause of that
dangerous suppuration of wounds which so often
defeated the beneficent efforts of the surgeon and
made an operation in a hospital ward as dangerous to
the patient as residence in a plague-stricken city.
Yet this is the result which the assiduous studies of
the biologists, provided with laboratories and maintenance by Continental States, have in due time brought
to light.   Theodore Schwann, professor at Liége, first
showed that these Bacteria are the cause of the putrefaction of organic substances, and subsequently the
French chemist Pasteur, professor in the École Normale
of Paris, confirmed and extended Schwann's discovery,
so as to establish the belief that all putrefactive
changes are due to such minute organisms, and that
if these organisms can be kept at bay no putrefaction
can occur in any given substance.

It was reserved for our countryman, Joseph Lister,
to apply this result to the treatment of wounds, and
by his famous antiseptic method to destroy by means
of special poisons the putrefactive organisms which
necessarily find their way into the neighbourhood of
a wound, or of the surgeon's knife and dressings, and

to ward off by similar means the access of such
organisms to the wounded surface. The amount of
death, not to speak of the suffering short of death,
which the knowledge of Bacteria gained by the microscope has thus averted is incalculable.

Yet further, the discoveries of Ehrenberg, Schwann,
and Pasteur are bearing fruit of a similar kind in other
directions. It seems in the highest degree probable
that the terrible scourge known as tubercular consumption or phthisis is due to a parasitic Bacterium (Bacillus),
discovered two years since by Koch of Berlin, as the
immediate result of investigations which he was com-
missioned to carry on at the public expense, in the
specially erected Laboratory of Public Health, by the
German Imperial Government. The diseases known
as erysipelas and glanders or farcy have similarly,
within the past few months in German State-supported
laboratories, been shown to be due to the attacks of
special kinds of Bacteria. At present this knowledge
has not led to a successful method of combating those
diseases, but we can hardly doubt that it will ultimately
do so. We are warranted in this belief by the fact
that the disease known as " splenic fever " in cattle and
" malignant pustule " or anthrax in man has likewise
been shown to be due to the action of a special kind
of Bacterium, and that this knowledge has, in the
hands of MM. Toussaint and Pasteur, led to a treatment
in relation to this disease similar to that of vaccination
in relation to small-pox. By cultivation a modified
growth of the anthrax parasite is obtained, which is
then used in order to inoculate cattle and sheep with

a mild form of the disease, such inoculation having
the result of rendering the cattle and sheep free from
the attacks of the severe form of disease, just as
vaccination or inoculation with cow-pox protects man
from the attack of the deadly small-pox.   One other
case I may call to mind in which knowledge of the
presence of Bacteria as the cause of disease has led to
successful curative treatment.   A not uncommon
affliction is inflammation of the bladder accompanied
by ammoniacal decomposition of the urine.   Microscopical investigation has shown that this ammoniacal
decomposition is entirely due to the activity of a
Bacterium.   Fortunately this Bacterium is at once
killed by weak solutions of quinine, which can be
injected into the bladder without causing any injury
or irritation.   This example appears to have great
importance, because it is the fact that many kinds of
Bacteria are not killed by solutions of quinine, but
require other and much more irritant poisons to destroy
their life, which could not be injected into the bladder
without causing disastrous effects.   Since some
Bacteria are killed by one poison and some by
another, it becomes a matter of the keenest interest
to find out all such poisons ; and possibly among
them may be some which can be applied so as to
kill the Bacteria which produce phthisis, erysipelas,
glanders, anthrax, and other scourges of humanity,
whilst not acting injuriously upon the body of
the victim in which these infinitesimal parasites
are doing their deadly work.   In such ways as
this biology has turned the toy "magnifying-glass"

of the last century into a saver of life and health. (See Appendix A.)

No less has the same agency revolutionised the thoughts of men in every branch of philosophy and speculation. The knowledge of the growth of the chick from the egg and of other organisms from similarly constituted beginnings has been slowly and continuously gained by prodigious labour, extending over generation after generation of students who have occupied the laboratories and lived on the stipends provided by the Governments of European States— not English, but chiefly German. It is this history of the development of the individual animal and plant from a simple homogeneous beginning to a complex heterogeneous adult which has furnished the starting-point for the wide-reaching Doctrine of Evolution. It is this knowledge, coupled with the knowledge of the myriad details of structure of all kinds of animals and plants which the faithful occupants of laboratories and the guardians of biological collections have in the past hundred years laboriously searched out and recorded—it is this which enabled Darwin to propound, to test, and to firmly establish his theory of the origin of species by natural selection, and finally to bring the origin, development, and progress of man also into the area of physical science. I have said enough, in referring only to two very diverse examples of the far-reaching consequences flowing from the discoveries of single-minded investigators in biological science, to remind my hearers that in the domain of biology, as in other sciences, the results attained by those who have laboured simply to extend our knowledge of the structure and properties of living things, in the faith that every increase of knowledge will ultimately bring its blessing to humanity, have in fact led with astonishing rapidity to conclusions affecting most profoundly both the bodily and the mental welfare of the community.

We who know the beneficent results which must flow more and more from the labours of those who are able to create new knowledge of living things, or, in other words, are able to aid in the growth of biological science, must feel something more than regret —even indignation—that England should do so small a proportion of the laborious investigation which is necessary, and is being carried on for our profit by other nationalities. It must not be supposed, because we have had our Harvey and our Darwin, our Hunter and our Lister, that therefore we have done and are doing all that is needful in the increase of biological science. The position of this country in relation to the progress of science is not to be decided by the citation of great names.

We require to look more fully into the matter than this. The question is not whether England has produced some great discoverers, or as many as any other nationality, but whether we might not, with advantage to our own community and that of the civilised world generally, do far more in the field of scientific investigation than we do.

It may be laid down as a general proposition, to which I know of no important exception, that scientific discovery has only been made by one of two classes of men, namely—(1) those whose time could be devoted to it in virtue of their possessing inherited fortunes; (2) those whose time could be devoted to it in virtue of their possessing a stipend or endowment especially assigned to them for that purpose.

Now it is a very remarkable fact that in England, far more than in any other country, the possessors of private fortunes have devoted themselves to scientific investigation. Not only have we in all parts of the country numerous *dilettanti*[1] who, especially in various branches of biology, do valuable work in continually adding to knowledge, quietly pursuing their favourite study without seeking to reach to any great eminence, but it is the fact that many of the greatest names of English discoverers in science are those of men who held no professional position designed to maintain an investigator, but owed their opportunity simply to the fact that they enjoyed a more or less ample income by inheritance. Thus, Harvey possessed a private fortune, Darwin also, and Lyell. Such also is true of some of the English naturalists, who more recently have most successfully devoted their energies to research.[2] Those who wish to defend the present neglect of the Government and of public institutions to provide means for the carrying on of scientific research in

---

[1] I use this word in its best and truest sense, and would refer those who have been accustomed to associate with it some implication of contempt, to the wise and appreciative remarks of Goethe on Dilettanti.

[2] I had in mind my friend Professor Frank Balfour of Cambridge, whose untimely death has been an irreparable blow to the progress of science in this country. December 1889.

this country are accustomed to declare as a justifica-
tion for this neglect that we do very well without such
provision, inasmuch as the cultivation of science here
flourishes in the hands of those who are in a position
of pecuniary independence. The reply to this is obvious. If those few of our countrymen who by accident
are placed in an independent position show such ability
in the prosecution of scientific research, how much
more would be effected in the same direction were the
machinery provided to enable those also who are *not*
accidentally favoured by fortune to enter upon the same
kind of work? The number of wealthy men who have
distinguished themselves in scientific research in England is simply evidence that there is a natural ability
and liking for such work in the English character, and
is a distinct encouragement to those who have it in
their power to do so to offer the opportunity of devoting themselves to research to a larger number of the
members of the community. It is impossible to doubt
that there are hundreds of men amongst us who have
as great capacity for scientific discovery as those whom
fortune has favoured with leisure and opportunity.
It cannot be doubted that, were the means
provided to enable even a proportion of such
men to give themselves up to scientific investigation, great discoveries, of no less importance to
the world than those relative to the causes of
disease and the development of living things
from the egg which I have cited, would be
made as a direct consequence of their activity,
whereas now we must wait until in due course

of time these discoveries shall be made for us in the laboratories of Germany, France, or Russia.

It should further be pointed out that it is altogether a mistake to suppose that the existence amongst us of a few very eminent men is any evidence that we are contributing largely to the hard work of careful study and observation which really forms the material upon which the conclusions of eminent discoverers are based. You will find in every department of biological knowledge that the hard work of investigation is being carried on by the well-trained army of German observers. Whether you ask the zoologist, the botanist, the physiologist, or the anthropologist, you will get the same answer : it is to German sources that he looks for new information ; it is in German workshops that discoveries, each small in itself, but gradually leading up to great conclusions, are daily being made. To a very large extent the business of those who are occupied with teaching or applying biological science in this country consists in making known what has been done in German laboratories ; our English students flock to Germany to learn the methods of scientific research ; and to such a state of weakness is English science reduced for want of proper nurture and support, that even on some of the rare occasions when a capable investigator of biological problems has been required for the public service, it has been necessary to obtain the assistance of a foreigner trained in the laboratories of Germany. (See Appendix B.)

Let me now briefly explain what are the arrangements, in number and in kind, which exist in other countries for the purpose of promoting the advancement of biological science, and are wanting in this country.

In the German Empire, with a population of 45,000,000, there are twenty-one universities. These universities are very different from anything which goes by the name in this country. Amongst its other arrangements devoted to the study and teaching of all branches of learning and science, each university has five institutes, or establishments, devoted to the prosecution of researches in biological science. These are respectively the physiological, the zoological, the anatomical, the pathological, and the botanical. In one of these universities of average size, each of the institutes named consists of a spacious building containing many rooms fitted as workshops, provided with instruments, a museum, and, in the last instance, with an experimental garden. All this is provided and maintained by the State. At the head of each institute is the university professor respectively of physiology, of zoology, of anatomy, of pathology, or of botany. He is paid a stipend by the State, which in the smallest university is as low as £120, but may be in others as much as £700, and averages say £400 a year. Considering the relative expenditure of the professional classes in the two countries, this average may be taken as equal to £800 a year in England.[1]

---

[1] From the fact that the salaries of judges, civil servants, military and naval officers, parsons and schoolmasters, as also the fees of physicians and lawyers, are in Germany even less than half what is paid to

Besides the professor, each institute has attached to it, with salaries paid by the State, two qualified assistants, who in course of time will succeed to independent positions. A liberal allowance is also made to each institute by the State for the purchase of instruments, material for study, and for the pay of servants, so that the total expenditure on professor, assistants, laboratory service, and maintenance, averages £800 a year for each institute—reaching as much as £2000 or £3000 a year in the larger universities. It is the business of the professor, in conjunction with his assistants and the advanced students, who are admitted to work in the laboratories free of charge, to carry on investigations, *to create new knowledge* in the several domains of physiology, zoology, anatomy, pathology, and botany. It is for this that the professor receives his stipend, and it is on his success in this field of labour that his promotion to a more important or better paid post in another university depends. In addition to and irrespectively of this part of his duties, each professor is charged with the delivery of courses of lectures and of elementary instruction to the general students of the university, and for this he is allowed to charge a certain fee to each student, which he receives himself; the total of such fees may, in the case of a largely attended university and a popular subject, form a very important addition to the professorial income; but it is distinctly to be understood that such payment by fees is only an *addition* to the professor's

their representatives in England, I think that we are justified in making this estimate.

income, quite independent of his stipend and of his
regular occupation in the laboratory : it is paid from
a separate source and for a separate object. There are
thus in the German Empire more than 100 such institutes devoted to the prosecution of biological discovery, carried on at an annual cost to the State of
about £80,000, equal to about £160,000 in England,
providing posts of graduated value for 300 investigators, some of small value, sufficient to carry the young
student through the earlier portion of his career,
whilst he is being trained and acting as the assistant
of more experienced men—others forming the sufficient
but not too valuable prizes which are the rewards of
continuous and successful labour.

In addition to these university institutes, there
are in Germany such special laboratories of research,
with duly salaried staff of investigators, as the Imperial
Sanitary Institute of Berlin, and the large museums
of Berlin, Bremen, and other large towns, corresponding to our own British Museum of Natural History.

Moreover, we must be careful to note, in making
any comparison with the arrangements existing in
England, that there are, in addition to the universities
in Germany, a number of other educational institutions, at least equal in number, which are known as
polytechnic schools, technical colleges, and agricultural
colleges.   These furnish posts of emolument to a
limited number of biological students, who give
courses of instruction to their pupils, but they have
not the same arrangements for research as the universities, and are closely similar to those colleges

which have been founded of late years in the provincial towns of England, such as Bristol, Nottingham,
and Leeds.   The latter are sometimes quoted by
sanguine persons, who are satisfied with the neglected
condition of scientific training and research in this
country, as really sufficient and adequate representatives of the German universities.   As a matter of
fact, the excellent English colleges in question do not
present anything at all comparable to the arrangements of a German university, and are, in respect of
the amount of money which is expended upon them,
the number of their teaching staff, and the efficiency
of their laboratories, inferior not merely to the smallest
German university, but inferior to many of the technical schools of that country.

Passing from Germany, I would now ask your
attention for a moment to an institution which is
supported by the French Government, and which—
quite irrespective of the French university system,
which is not on the whole superior to our own—
constitutes one of the most effective arrangements in
any European State for the production of new knowledge.   The institution to which I allude is the Collège
de France in Paris—coexisting there with the Sor-
bonne, the École de Médecine, the École Normale, the
Jardin des Plantes, and other State-supported institutions—in which opportunity is provided for those
Frenchmen who have the requisite talent to pursue
scientific discovery in the department of biology,
and in other branches of science.   I particularly mention the Collège de France, because it appears to me

that the foundation of such a college in London would
be one of the simplest and most direct steps that could
be taken towards filling, in some degree, the void
from which English science suffers. The Collège de
France is divided into a literary and a scientific faculty.
Each faculty consists of some twenty professors.
Each professor in the scientific faculty is provided
with a laboratory and assistants (as many as four
assistants in some cases), and with a considerable
allowance for the expenses of the instruments and
materials required in research. The personal stipend
of each professor is £400, which has been increased
by an additional £100 a year in some cases from the
Government Department charged with the promotion
of higher studies. The professors in this institution,
as in the German universities, when a vacancy occurs,
have the right of nominating their future colleague,
their recommendation being accepted by the Government. The professors are not expected to give any
elementary instruction, but are directed to carry on
original investigations, in prosecuting which they may
associate with themselves pupils who are sufficiently
advanced to join in such work; and it is further the
duty of each professor to give a course of forty lectures
in each year upon the results of the researches in
which he is engaged. There are at present among
the professors of the Collège de France four of the
most distinguished among contemporary students of
biological science : Professor Brown-Séquard, Professor
Marey, Professor Balbiani, and Professor Ranvier.
Every one who is acquainted with the progress of

discovery in physiology, minute anatomy, and embryology, will admit that the opportunities afforded to these men have not been wasted; they have, as the result of the position in which they have been placed, produced abundant and most valuable work, and have, in addition, trained younger men to carry on the same line of activity. It was here, too, in the Collège de France, that the great genius of Claude Bernard found the necessary conditions for its development.

Let us now see how many and what kind of institutions there are in England devised so as to promote the making of new knowledge in biological science. Most persons are apt to be deceived in this matter by the fact that the terms "university," "professorship," and "college" are used very freely in England in reference to institutions which have no pecuniary resources whatever, and which, instead of corresponding to the German arrangements which go by these names, are empty titles, neither backed by adequate subsidy of the State nor by endowment from private sources.

In England, with its 25,000,000 inhabitants, there are only four universities which possess endowments and professoriates—viz. Oxford, Cambridge, Durham, and the Victoria (Owens College). Besides these, which are variously and specially organised each in its own way, there are the London Colleges (University and King's), the Normal School of Science at South Kensington, and various provincial colleges, which are to a small and varying extent in possession of funds which could be or are used to promote scientific research. Amongst all these variously arranged institutions there is an extraordinarily small amount of provision for biological research. In London there is one professorship only, that at the Normal School of Science, which is maintained by a stipend paid by the State, and has a laboratory and salaried assistants, similarly maintained, in connection with it. The only other posts in London which are provided with stipends intended to enable their holders to pursue researches in the domain of biological science, are the two chairs of physiology and of zoology at University College, which, through the munificence of a private individual (Mr. Jodrell), have been endowed to the extent of £250 a year each. (See Appendix C.) To these should be added, in our calculation, certain posts in connection with the British Museum of Natural History and the Royal Gardens at Kew, maintained by the State; though it must be remembered that a large part of the expenditure in those institutions is necessarily taken up in the preservation of great national collections, and is not applicable to the subvention of investigators. We may, however, reckon about six posts, great and small, in the British Museum, and four at Kew, as coming into the category which we have in view. In London, then, we may reckon approximately some fourteen or fifteen subsidised posts for biological research. In Oxford there fall under this category the professorship of anatomy and his assistant, that of physiology, that of zoology, that of botany. The Oxford professorships are well

supported by endowment, averaging £700 or £800 a
year; but they are inadequately provided with assistance as compared with corresponding German positions. Whilst Oxford has thus five posts, Cambridge
has at present the same number, though the stipends
are of less average value. In regard to Durham, it
does not appear that the biological professorships
(which have their seat in the Newcastle College of
Science) are supported by stipends derived from
endowment; they fall under another category, to
which allusion will be made below, of purely teaching
positions, supported by the fees paid for such teaching
by pupils. The Victoria University (Owens College,
Manchester), supports its professors of physiology,
anatomy, zoology, botany, and pathology, by means
partly of endowment, partly of pupils' fees. By the
provision of adequate laboratories and of salaries for
assistants to each professor, and of student-fellow-
ships, Owens College gives direct support to original
investigation. We may reckon five major and eight
minor posts as dedicated to biological research in this
college. Altogether, then, we have 15 positions in
London and 23 in the provinces (taking assistantships,
and professorships, and curatorships together)—a total
of 38 in all England with its 25,000,000 inhabitants,
as against the 300 in Germany with its 45,000,000
inhabitants. In proportion to its population (leaving
aside the consideration of its greater wealth), England
has only about one-fourth of the provision for the
advancement of biological research which exists in
Germany.

It would not be fair to reckon in this comparison the various biological professorships in small colleges recently created, and paid to a small extent by stipends derived from endowments, in the provincial towns of England, for the holders of these chairs are called upon to teach a variety of subjects, for instance, zoology, botany, and geology combined; and not only is the devotion of the energies of their teaching staff to scientific discovery not contemplated in the arrangement of these institutions, but, as a matter of fact, the large demands made on the professors in the way of teaching must deprive them of the time necessary for any serious investigation.   Such posts, in the fact that neither time, assistants, nor proper laboratories are provided to enable their holders to engage in scientific research, are schoolmasterships rather than professorships, as the word is used in German universities. (See Appendix C.)

One result of the exceedingly small provision of positions in England similar to those furnished by the German university system, and of the irregular, uncertain character of many of those which do exist, is that there is an insufficient supply of young men willing to enter upon the career of zoologist, botanist, physiologist, or pathologist as a profession.   The number of posts is too small to create a profession, i.e. an avenue of success; and consequently, whereas in Germany there is always a large body of new men ready to fill up the vacancies as they occur in the professorial organisation, in England it very naturally does not appear to our university students as a reasonable thing to enter upon research as a profession, when the chances of employment are so few and far between.

Before stating, as I propose to do, what appears to me a reasonable and proper method of removing to some extent the defect in our national life due to the want of provision for scientific research, I will endeavour to meet some of the objections which are usually raised to such views as those which I am advocating. The endowment of research by the State, or from public funds of any kind, is opposed on various grounds. One is that such action on the part of the Government is well enough in Continental States, but is contrary to the spirit of English statecraft, which leaves scientific as well as other *enterprise* to the individual initiative of the people. This objection is based on error, both as to fact and theory. It is well enough to leave to individual effort the conduct of such enterprises as are remunerative to the parties who conduct them; but it is a mistake to speak of scientific research as an "enterprise" at all. The mistake arises from the extraordinary pertinacity with which so-called "invention" is confounded with the discovery of scientific truth. New knowledge in biological or other branches of science cannot be sold; it has no marketable value. Koch could not have sold the discovery of the Bacterium of phthisis for as much as sixpence, had he wished to do so. Accordingly, we find that there is not, and never has been, any tendency among the citizens of this country to provide for themselves institutions for the manufacture of an article of so little pecuniary value to the individual who turns it out as is new knowledge. On the other hand, as a matter of fact, the providing of means for the manufacture of that article is not only not foreign to English statecraft, but is largely, though not largely enough, undertaken by the English State. The Royal Observatories, the British Museum, the Royal Gardens at Kew, the Geological Survey, the Government grant of £4000 a year to the Royal Society, the £300 or £400 a year (not a large sum) expended through the medical officer of the Privy Council upon the experimental investigation of disease, are ample evidence that such providing of means for creating new knowledge forms part of the natural and recognised responsibilities of the British Government. Such a responsibility clearly is recognised in this country, and does fall, according to the present arrangement of things, upon the central Government. What we have to regret is, that those who temporarily hold the reins of government fail to perceive the lamentable inadequacy of the mode in which this responsibility is met.

A second objection which is made to the endowment of research by public funds, or by other means, such as voluntary contributions, is this : it is stated that men engaged in scientific research ought to *teach*, and thus gain their livelihood. It is argued, in fact, that there is no need whatever to provide stipends or laboratories for researchers, since they have only to stand up and teach in order to make incomes sufficient to keep them and their families, and to provide themselves with laboratories. This is a very plausible statement, because it is the fact that some investigators have also been excellent lecturers, and have been able to make an income by teaching whilst carrying on a limited amount of scientific investigation. But neither by teaching in the form of popular lectures, nor by teaching university or professional students who desire as a result to pass some examination test, is it possible, where there is a fair field and no favour, for a man to gain a reasonable income and at the same time to leave himself time and energy to carry on original investigations in science.

In some universities, such as those of Scotland, the privilege of conferring degrees of pecuniary value to their possessors becomes a source of income to the professors of the university ; they are, in fact, able to make considerable incomes, independently of endowment, by compelling the candidates for degrees to pay a fee to each professor in the faculty for the right of attending his lectures and of presentation to the degree. Consequently, teaching here appears to be producing an income which may support a researcher ; in reality, it is the acquisition of the university degree, and not necessarily the teaching, for which the pupil pays his fee. Where the teacher is unprotected by any compulsory regulations (such as that which requires attendance on his lectures and fee-payment on the part of the pupils) it is *impossible* for him to obtain such an income by teaching for one hour a day as will enable him to devote the rest of the day to unremunerative study and investigation, for the following reason. Other teachers, equally satisfactory as teachers, will enter into competition with him, without having the same intention of teaching for one hour only, and of carrying on researches for the rest of the day. They will contemplate teaching for six hours a day, and they will accordingly offer to those who require to be taught either six hours' teaching for the same fee which the researcher charges for one, or one hour for a sixth part of that fee. Consequently the unprotected researcher will find his lecture-room deserted—pupils will naturally go to the equally good teacher who gives more teaching for the same fee, or the same teaching for a less fee. And no one can say that this is not as it should be. The university pupil requires a certain course of instruction, which he ought to be able to buy at the cheapest rate. It does not seem to be doing justice to the pupil to compel him to form one of a class consisting of some hundreds of hearers, where he can obtain but little personal supervision or attention from the teacher, whereas if he had the free disposal of his fee, he might obtain six times the amount of attention from another teacher. This arrangement does not seem to be justifiable, even for the purpose of providing the university professor with an income and leisure to pursue scientific research. The student's fee should pay for a given amount of teaching at the market value, and he has just cause of complaint if, by compulsory enactments, he is taxed to provide the country with scientific investigation.

Teaching must, in all fairness, ultimately be paid for as teaching, and scientific research must be provided for, out of other funds than those extracted from the pockets of needy students, who have a reasonable right to demand, in return for their fees, a full modicum of instruction and direction in study.

In the German universities, the professor receives a stipend which provides for him as an investigator. He also gives lectures, for which he charges a fee, but no student is compelled to attend those lectures as a condition of obtaining his degree. Accordingly, independent teachers can, and do, compete with the professor in providing for the student's requirements in the matter of instruction. As a consequence, the fees charged for teaching are exceedingly small, and the student can feel assured that he is obtaining his money's worth for his money. He is not compelled to pay any fee to any teacher as a condition of his promotion to the university degree. In a German university, if the professor in a given subject is incompetent, or the class overcrowded, the student can take his fee to a private teacher, and get better teaching; all that is required of the candidate, as a condition of his promotion to the Doctor's degree, is that he shall satisfy the examination-tests imposed by the faculty, and produce an original thesis.

Unless there be some such compelling influence as that obtaining in the Scotch universities, enabling the would-be researcher to gather to him pupils and fees without fear of competition, it seems impossible that he should gain an income by teaching whilst reserving to himself time and energy for the pursuit of scientific inquiry. It is thus seen that the necessity of endowment, in some form or another, to make provision for scientific research, is a reality, in spite of the suggestion that teaching affords a means whereby the researcher may readily provide for himself. The simple fact is that a teacher can only make a sufficient income by teaching, on the condition that he devotes his whole time and energy to that occupation.

Whilst I feel called upon to emphatically distinguish the two functions—viz. that of *creating new knowledge*, and that of *distributing existing knowledge*— and to maintain that it is only by arbitrary and undesirable arrangements, not likely to be tolerated, or, at any rate, extended, at the present day, that the latter can be made to serve as the support of the former, I must be careful to point out that I agree most cordially with those who hold that it is an excellent thing for a man who is engaged in the one to give a certain amount of time to the other. It is a matter of experience that the best teachers of a subject are, *cæteris paribus*, those who are actually engaged in the advancement of that subject, and who have shown such a thorough understanding of that subject as is necessary for making new knowledge in connection with it. It is also, in most cases, a good thing for the man engaged in research to have a certain small amount of change of occupation, and to be called upon to take such a survey of the subject in connection with which his researches are made, as is involved in the delivery of a course of lectures and other details of teaching. Though it is not a thing to be contemplated that the researcher shall sell his instruction at a price sufficiently high to enable him to live by teaching, yet it is a good thing to make teaching an additional and subsidiary part of his life's work. This end is effected in Germany by making it a duty of the professor, already supported by a stipend, to give some five or six lectures a week during the academical session, for which he is paid by the fees of his hearers. The fees are low, but are sufficient to be an inducement; and, inasmuch as the attendance of the students is not compulsory, the professor is stimulated to produce good and effective lectures at a reasonable charge, so as to attract pupils who would seek instruction from some one else if the lectures were not good or the fees too high. Indeed, in Germany this system works so much to the advantage of the students, that the private teachers of the universities at one time obtained the creation of a regulation forbidding the professors to reduce their fees below a certain minimum, since, with so low a fee as some professors were charging, it was impossible for a private teacher to compete! This state of things may be compared, with much advantage, with the condition of British universities. In these we hear, from one direction, complaints of the high fees charged and of the ineffective teaching given by the professoriate; and in other universities, where no adequate fees are allowed to the professors as a stimulus to them to offer useful and efficient teaching, we find that the teaching has passed entirely out of their hands into those of college tutors and lecturers. The fact is that a satisfactory relation between teaching and research is one which will not naturally and spontaneously arrange itself.   It can hardly be said to exist in any British university or college, but the method has been thought out and carried into practice in Germany.   It consists in giving a competent researcher a stipend and a laboratory for his research work, and then requiring him to do a small amount of teaching, remunerated by fees proportionate to his ability and the pains which he may take in his teaching.   If you pay him a fixed sum as a teacher, or artificially insure the attendance of his class, instead of letting this part of his income vary simply and directly with the attractiveness of his teaching, you will find as the result that (with rare exceptions) he will not give effective and useful teaching.   He will naturally tend to do the minimum required of him, in a perfunctory way.   On the other hand, if you leave him without stipend as a researcher, dependent on the fees of pupils for an income, he will give all his time and energies to teaching, he will cease to do any research, and become, *pro tanto*, an inferior teacher.

A third objection which is sometimes made to the proposition that scientific research must be supported and paid for as such, is the following : It is believed by many persons that a man who occupies his best energies in scientific research can always, if he choose, make an income by writing popular books or newspaper articles in his spare hours ; and, accordingly, it is gravely maintained that there is no need to provide stipends and the means of carrying on their work for researchers.   To do so, according to this view, would be to encourage them in an exclusive reticence, and to remove from them the inducement to address the public on the subject of their researches, by which the public would lose valuable instruction.

This view has been seriously urged, or I should not here notice it. Any one who is acquainted with the sale of scientific books, and the profits which either author or publisher makes by them, knows that the suggestion which I have quoted is ludicrous. The writing of a good book is not a thing to be done in leisure moments, and such as have been the result of original research have cost their authors often years of labour apart from the mere writing. Mr. Darwin's books, no doubt, have had a large sale; but that is due to the fact, apart from the exceptional genius of the man who wrote them, that they represent some thirty or more years of hard work, during which he was silent. There is not a sufficiently large public interested in the progress of science to enable a researcher to gain an income by writing books, however great his literary facility. A school-book or class-book may now and then add more or less to the income of a scientific investigator; but he who becomes the popular exponent of scientific ideas, except in a very moderate and limited degree, must abandon the work of creating new knowledge. The professional *littérateur* of science is as much removed by his occupation from all opportunity of serious investigation as is the professional teacher who has to consume all his time in teaching. Any other pro- fession—such as the Bar, Medicine, or the Church—

is more likely to leave one of its followers time and
means for scientific research than is that of either the
popular writer or the successful teacher.

We have, then, seen that there is no escape from
the necessity of providing stipends and laboratories
for the purpose of creating new knowledge, as is done
in Continental States, if we are agreed that more of
this new knowledge is needed and is among the
products which a civilised community is bound to
turn out, both for its own benefit and for that of
the community of States, which give to and take from
one another in such matters.

There are some who would finally attack our
contention by denying that new knowledge is a good
thing, and by refusing to recognise any obligation on
the part of England to contribute her share to that
common stock of increasing knowledge by which she
necessarily profits.   Among such persons are those
who would prohibit altogether the pursuit of experimental physiology in England, and yet would not and
do not hesitate to avail themselves of the services
of medical men, whose power of rendering those
services depends on the fact that they have learnt the
results obtained by the experiments of physiologists
in other countries or in former times.   In reference to
this strange contempt and even hatred of science,
which undoubtedly has an existence among some
persons of consideration, even at the present day,
I shall have a few words to say before concluding
this address.   I have now to ask you to listen to
what seems to me to be the demand which we should

make, as members of a British Association for the
Advancement of Science, in respect of adequate
provision for the creation of new knowledge in the
field of biology in England.

Taking England alone, as distinct from Scotland
and Ireland, we require, in order to be approximately
on a level with Germany, forty new biological institutes, distributed among the five branches of physiology, zoology, anatomy, pathology, and botany—forty
in addition to the fifteen which we may reckon (taking
one place with another) as already existing. The
average cost of the buildings required would be about
£4000 for each, giving a total initial expenditure of
£160,000 ; the average cost of stipends for the director,
assistants, and maintenance we may calculate at £1500
annually for each, or £60,000 for the forty—equal to
a capital sum of £2,000,000. These institutes should
be distributed in groups of five—eight groups in all—
throughout the country. One such group would be
placed in London (which is, at present, almost totally
destitute of such arrangements), one in Bristol, one in
Birmingham, one in Nottingham, one in Leeds, one in
Newcastle, one in Ipswich, one in Cardiff, one in Plymouth—in fact, one in each of the great towns of the
kingdom where there is at present, or where there
might be with advantage, a centre of professional
education and higher study. The first and the most
liberally arranged of these biological institutes—embracing its five branches, each with its special laboratory
and staff—should be in London. If we can have
nothing else, surely we may demand, with some hope

that our request will eventually obtain compliance, the formation in London of a College of Scientific Research similar to that of Paris (the Collège de France). It is one of the misfortunes and disgraces of London that —alone amongst the capitals of Europe, with the exception of Constantinople—it is destitute of any institution corresponding to the universities and colleges of research which exist elsewhere.

Either in connection with a properly organised teaching university or as an independent institution, it seems to me a primary need of the day that the Government should establish in London laboratories for scientific research. Two hundred and fifty years ago Sir Thomas Gresham founded an institution for scientific research in the City of London. The property which he left for this purpose is now estimated to be worth three millions sterling. This property was deliberately appropriated to other uses by the Corporation of the City of London and the Mercers' Company about a hundred years since, with the consent of both Houses of Parliament. By this outrageous act of spoliation these Corporations, who were the trustees of Gresham, have incurred the curse which he quaintly inserted in his will in the hope of restraining them from attempts to divert his property from the uses to which he destined it. "Gresham's curse" runs as follows : " And that I do require and charge the said Corporations and chief governors thereof, with circumspect Diligence and without long Delay, to procure and see to be done and obtained, as they will answer the same before Almighty God ; (for if they or any of them should neglect the obtaining of such Licenses or Warrants, which I trust can not be difficult, nor so chargeable, but that the overplus of my Rents and Profits of the Premisses hereinbefore to them disposed, will soon recompense the same ; because to soe good Purpose in the Commonwealth, no Prince nor Council in any Age, will deny or defeat the same. And if conveniently by my Will or other Convenience, I might assure it, I would not leave it to be done after my death, then the same shall revert to my heirs, whereas I do mean the same to the Commonwealth, and then the Default thereof shall be to the Reproach and Condemnation of the said Corporations afore God)." I confess that I find it difficult to see how the present representatives of the Corporations who perverted Gresham's trust are to escape from justly deserving the curse pronounced against those Corporations, unless they conscientiously take steps to restore Gresham's money to its proper uses. Let us hope that Gresham's curse may be realised in no more deadly form than that of an Act of Parliament repealing the former one which sanctioned the perversion of Gresham's money. Such a sequel to the Report of the Commission which has recently inquired into the proceedings of the Corporation and Companies of the City of London is not unlikely.

Whilst we should, I think, especially press upon public attention the need for an institute of scientific research in London, and indicate the source from which its funds may be fitly derived, we must also urge the foundation of other institutes in the provinces upon the scale already sketched, because it is only by the existence of numerous posts, and of a series of such posts—some of greater and some of less value, the latter more numerous than the former—that anything like a professional career for scientific workers can be constructed. It is especially necessary to constitute what I have termed "assistantships," that is, junior posts in which younger men assist and are trained by more experienced men. Even in the few institutions which do already exist additional provision of this kind is what is wanted more than anything else, so that there may be a progressive career open to the young student, and a sufficient field of trained investigators from which to select in filling up the vacancies in more valuable positions.

I am well aware that it will be said that the scheme which I have proposed to you is gigantic and almost alarming in respect of the amount of money which it demands. One hundred and sixty thousand pounds a year for biology alone must seem, not to my hearers, but to those who regard biology as an amusing speculation— that is to say, who know little or nothing about it — an extravagant suggestion. Unfortunately it is also true that such persons are very numerous —in fact, constitute an overwhelming majority of the community; but they are becoming less numerous every day. The time will come, it seems possible, when there will be more than one member of the Government who will understand and appreciate the value of scientific research.

There are already a few members of the House
of Commons who are fully alive to its significance
and importance.

We may have to wait for the expenditure of such
a sum as I have named, and possibly it may be derived
ultimately from local rather than imperial sources,
though I do not see why it should be ; yet I think it
is a good thing to realise now that this is what we
ought to expend in order to be on a level with Germany.
This apparently extravagant and unheard-of appropri-
ation of public money is actually made every year in
Germany.

I think it is well to put the matter before you in
this definite manner, because I have reason to believe
that even those whom we might expect to be well
informed in regard to such matters, are not so, and as
a consequence there is not that keen sense of the inferiority and inadequacy of English arrangements in
these matters which one would gladly see actuating
the conduct of English statesmen.   For instance, only
a few years ago, when speaking at Nottingham, the
present Prime Minister, who has taken an active part
in rearranging our universities, and has, it is well
known, much interest in science and learning, stated
that £27,000, the capital sum expended on the Nottingham College of Science, was a very important contribution to the support of learning in this country,
amounting, as he said he was able to state, from the
perusal of official documents, to as much as one-third
of what was spent in Germany during the past year
upon her numerous universities, which were so often

held up to England as an example of a well supported academical system. Now, I do not think that Mr. Gladstone can have ever had the opportunity of considering the actual facts with regard to German universities, for he was in this instance misled by the official return of expenditure on a single university, namely, that of Strasburg ; the total annual expenditure on the twenty-one German universities being, in reality, about £800,000, by the side of which a capital sum of £27,000 looks very small indeed. I cannot but believe that if the facts were known to public men, in reference to the expenditure incurred by foreign States in support of scientific inquiry, they would be willing to do something in this country of a sufficient and statesmanlike character. As it is, the concessions which have been made in this direction appear to me to be in some instances not based upon a really comprehensive knowledge of the situation. Thus the tentative grant of £4000 a year from the Treasury to the Royal Society of London appears to me not to be a well-devised experiment in the promotion of scientific research by means of grants of money, because it is on too small a scale to produce any definite effect, and because the money cannot be relied upon from year to year as a permanent source of support to any serious undertaking.

The Royal Society most laboriously and conscientiously does its best to use this money to the satisfaction of the country, but the task thus assigned to it is one of almost insurmountable difficulty. In fact, no such miniature experiments are needed. The experiment has been made on a large scale in Germany, and satisfactory results have been obtained. The reasonable course to pursue is to benefit by the experience, as to details and methods of administration, obtained in the course of the last sixty years in Germany, and to apply that experience to our own case.

It is quite clear that " the voluntary principle " can do little towards the adequate endowment of scientific research. Ancient endowments belonging to the country must be applied thereto, or else local or imperial taxes must be the source of the necessary support. Seeing that the results of research are distinctly of imperial, and not of local value—it would seem appropriate that a portion of the imperial revenue should be devoted to their achievement. In fact, as I have before mentioned, the principle of such an application of public money has long been admitted, and is in operation. (See Appendix C.)

Whilst voluntary donations on the part of private persons can do little to constitute a fund which shall provide the requisite endowment for the scheme of biological institutes which I have sketched (not to mention those required for other branches of science), yet those who are interested in the progress of scientific investigation may by individual effort do something, however little, towards placing research in a more advantageous position in this country. Supposing it were possible, as I am sanguine enough to believe that it is, to collect in the course of a year or two from private sources a sum of £20,000 for the

maintenance of a biological laboratory and staff, it
would be necessary, in expending so limited a sum, to
aim at the provision of something which would be
likely to produce the largest and most obvious results
in return for the outlay, and to benefit the largest
number of scientific observers in this department.

I believe that it is the general opinion among
biologists that there could be no more generally
useful institution thus set in operation than a biological laboratory upon the sea-coast, which, besides
its own permanent staff of officers, would throw open
its resources to such naturalists as might from time
to time be able to devote themselves to researches
within its precincts.   There is no such laboratory on
the whole of the long line of British coast.   At
Naples there is Dr. Dohrn's celebrated and invaluable
laboratory, which is frequented by naturalists from all
parts of the world ; at Trieste the Austrian Government supports such a laboratory ; at Concarneau,
Roscoff, and Villefranche, the French Government has
such institutions ; at Beaufort, in North Carolina,
the Johns Hopkins University has its marine laboratory ; and at Newport, Professor Alexander Agassiz
has arranged a very perfect institution also for the
study of marine life.   In spite of the great interest
which English naturalists have always taken in the
exploration of the sea and marine organisms—
in spite of the fact that the success and even
the existence of our fisheries-industries to a large
extent depends upon our gaining the knowledge
which a well-organised laboratory of marine biology

would help us to gain, there is actually no such institution in existence.

This is not the occasion on which to explain precisely how and to what extent a laboratory of marine zoology might be of national importance. I hope to see that matter brought before the Section during the course of our meeting. But I may point out now, that though it appears to me that the great need for biological institutes, to which I have drawn your attention, can *not* be met by private munificence, and must in the end be arranged for by the continued action of the Government in carrying out a policy to which it has for many years been committed, and which has been approved by Conservatives and Liberals alike—yet such a special institution as a laboratory of marine biology, serving as a temporary workshop to any and all of our numerous students of the important problems connected with the life of marine plants and animals, might very well be undertaken from private funds. Should it be possible, on the occasion of this meeting of the British Association in Southport, to obtain some promise of assistance towards the realisation of this project, I think we shall be able to congratulate ourselves on having done something, though small perhaps in amount, towards making better provision for biological research, and therefore something towards the advancement of science.[1]

---

[1] The wish here expressed was subsequently realised. By public subscription and Government aid, a well-equipped laboratory has been erected on the shore of Plymouth Sound. For an account of

In conclusion, let me say that, in advocating to-day the claim of biological science to a far greater measure of support than it receives at present from the public funds, I have endeavoured to press that claim chiefly on the ground of the obvious utility to the community of that kind of knowledge which is called biology. I have endeavoured to meet the opposition of those who object to the interference of the State wherever it may be possible to attain the end in view without such interference, but who profess themselves willing to see public money expended in promoting objects which are of real importance to the country, and which cannot be trusted to the voluntary enterprise arising from the operation of the laws of self-preservation and the struggle for wealth. There are, however, it seems to me, further reasons for desiring a thorough and practical recognition by the State of the value of scientific research. There are not wanting persons of some cultivation who have perceived and fully realised the value of that knowledge which is called science, and of its methods, and yet are anxious to restrain rather than to aid the growth of that knowledge. They find in science something inimical to their own interests, and accordingly either condemn it as dangerous and untrustworthy, or encourage themselves to treat it with contempt by asserting that "after all, science counts for very little"—a statement which is unhappily true in one sense, though totally untrue when it is intended

its history and organisation, the reader is referred to the appendix to No. V. of the present series of papers. December 1889.

to signify that the progress of science is not a matter which profoundly influences every factor in the well-being of the community. Amongst such people there is a positive hatred of science, which finds expression in their exclusion of it, even at this day, from the ordinary curriculum of public school education, and in the baseless though oft-repeated calumny that science is hostile to art, and is responsible for all that is harsh, ugly, and repulsive in modern life. To such opponents of the advancement of science, it is of little use to offer explanations and arguments. But we may, when we reflect on their instinctive hostility and the misrepresentations of science and the scientific spirit which it leads them to disseminate, console ourselves by bringing to mind what science really is, and what truly is the nature of that calling in which a man who makes new knowledge is engaged.

They mock at the botanist as a pedant, and the zoologist as a monomaniac; they execrate the physiologist as a monster of cruelty, and brand the geologist as a blasphemer; chemistry is held responsible for the abomination of aniline dyes and the pollution of rivers, and physics for the dirt and misery of great factory towns. By these unbelievers science is declared responsible for individual eccentricities of character, as well as for the sins of the commercial utilisers of new knowledge. The pursuit of science is said to produce a dearth of imagination, incapability of enjoying the beauty either of nature or of art, scorn of literary culture, arrogance, irreverence, vanity, and the ambition of personal glorification.

Such are the charges from time to time made by those who dislike science, and for such reasons they would withhold, and persuade others to withhold, the fair measure of support for scientific research which this country owes to the community of civilised States. Not in reply to these misrepresentations, but by way of contrast, I would here state what science seems to be to those who are on the other side, and how, therefore, it seems to them wrong to delay in doing all that the wealth and power of the State can do to promote its progress.

Science is not a name applicable to any one branch of knowledge, but includes all knowledge which is of a certain order or scale of completeness. All knowledge which is deep enough to touch the causes of things, is Science ; all inquiry into the causes of things is scientific inquiry. It is not only coextensive with the area of human knowledge, but no branch of it can advance far without reacting upon other branches ; no department of Science can be neglected without sooner or later causing a check to other departments. No man can truly say this branch of Science is useful and shall be cultivated, whilst this is worthless and shall be let alone ; for all are necessary, and one grows by the aid of another, and in turn furnishes methods and results assisting in the progress of that from which it lately borrowed.

We desire the increase and the support and the acceptance of Science, not only because it has a certain material value and enables men to battle with the forces of nature and to turn them to account, so as to increase both the intensity and the extension of healthy human life; that is a good reason, and for some persons, it may be, the only reason. But there is something to be said beyond this.

The pursuit of scientific discovery, the making of new knowledge, gratifies an appetite which, from whatever cause it may arise, is deeply seated in man's nature, and indeed is the most distinctive of his properties. Man owes this intense desire to know the nature of things, smothered though it often be by other cravings which he shares with the brutes, to an inherited race-perception stronger than the reasoning faculty of the individual. When once aroused and in a measure gratified, this desire becomes a guiding passion. The instinctive tendency to search out the causes of things, gradually strengthening as generation after generation of men have stumbled and struggled in ignorance, has at last become an active and widely extending force; it has given rise to a new faith.

To obey this instinct—that is, to aid in the production of new knowledge—is the keenest and the purest pleasure of which man is capable, greater than that derived from the exercise of his animal faculties, in proportion as man's mind is something greater and further developed than the mind of brutes. It is in itself an unmixed good, the one thing which commends itself as still "worth while" when all other employments and delights prove themselves stale and unprofitable.

Arrogant and foolish as those men have appeared

who, in times of persecution and in the midst of a contemptuous society, have, with an ardour proportioned to the prevailing neglect, pursued some special line of scientific inquiry, it is nevertheless true that in itself, apart from special social conditions, Science must develop in a community which honours and desires it before all things, qualities and characteristics which are the highest, the most human of human attributes. These are, firstly, the fearless love and unflinching acceptance of truth; hopeful patience; that true humility which is content not to know what cannot be known, yet labours and waits; love of Nature, who is not less, but more, worshipped by those who know her best; love of the human brotherhood for whom and with whom the growth of Science is desired and effected.

No one can trace the limits of Science, nor the possibilities of happiness both of mind and body which it may bring in the future to mankind. Boundless though the prospect is, yet the minutest contribution to the onward growth has its absolute and unassailable value; once made it can never be lost; its effect is for ever in the history of man.

Arts perish, and the noblest works which artists give to the world. Art (though the source of great and noble delights) cannot create nor perpetuate; it embodies only that which already exists in human experience, whilst the results of its highest flights are doomed to decay and sterility. A vain regret, a constant effort to emulate or to imitate the past, is the fitting and laudable characteristic of Art at the present day. There is, indeed, no truth in the popular partition of human affairs between Science and Art as between two antagonistic or even comparable interests ; but the contrast which they present in points such as those just mentioned is forcible. Science is essentially creative ; new knowledge—the experience and understanding of things which were previously non-existent for man's intelligence, is its constant achievement. And these creations never perish ; the new is built on and incorporates the old ; there is no turning back to recover what has lapsed through age ; the oldest discovery is even fresher than the new, yielding in ever increasing number new results, in which it is itself reproduced and perpetuated, as the parent in the child.

This, then, is the faith which has taken shape in proportion as the innate desire of man for more knowledge has asserted itself—namely, that there is no greater good than the increase of Science ; that through it all other good will follow.   Good as Science is in itself, the desire and search for it is even better, raising men above vile things and worthless competitions to a fuller life and keener enjoyments.   Through it we believe that man will be saved from misery and degradation, not merely acquiring new material powers, but learning to use and to guide his life with understanding.   Through Science he will be freed from the fetters of superstition ; through faith in Science he will acquire a new and enduring delight in the exercise of his capacities ; he will gain a zest

and interest in life such as the present phase of
culture fails to supply.

In opposition to the view that the pursuit of
Science can obtain a strong hold upon human life, it
may be argued, that on no reasonable ground can it
appear a necessary or advantageous thing to the
individual man to concern himself with the growth
and progress of that which is merely likely to benefit
the distant posterity of the human race. Our reply
is : Let those who contend for the reasonableness of
human motives develop, if they can, any theory of
human conduct in which reasonable self-interest shall
be man's guide. We do not contend for any such
theory. By reasoning we may explain and trace the
development of human nature, but we cannot change
it by any such process. It is demonstrably unreasonable for the individual man, guided by self-interest,
to share the dangers and privations of his brother-
man, and yet, in common with many lower animals,
he has an inherited quality which makes it a pleasure
to him to do so ; it is unreasonable for the mother to
protect her offspring, and yet it is the natural and
inherited quality of mothers to derive pleasure from
doing so ; it is unreasonable for the half-starved poor
to aid their wholly starving brethren, and yet such
compassion is natural and pleasurable to those who
show it, and is the constant rule of life. Unreasonable though these things are from the point of view
of individual self-interest, yet they are done because
to do them is pleasurable, to leave them undone a
pain. The race has, as it were, in these respects

befooled the individual, and in the course of evolution
has planted in him, in its own interests, an irrational
capacity for taking pleasure in doing that which no
reasoning in regard to self-interest could justify. As
with these lower and more widely distributed instincts, shared by man with some lower social animals,
so is it with this higher and more peculiar instinct—
the tendency to pursue new knowledge. Whether
reasonable or not, it has by the laws of heredity and
selection become part of us and exists ; its operation is
beneficial to the race ; its gratification is a source of
keen pleasure to the individual—an end in itself.
We may safely count upon it as a factor in human
nature ; it is in our power to cultivate and develop it,
or, on the other hand, to starve and distort it for a
while, though to do so is to waste time in opposing
the irresistible.

As day by day the old-fashioned stimulus to the
higher life loses the dread control which it once exercised over the thoughts of men, the pursuit of wealth
and the indulgence in fruitless gratifications of sense
become to an increasing number the chief concerns of
their mental life. Such occupations fail to satisfy
the deep desires of humanity ; they become wearisome
and meaningless, so that we hear men questioning
whether life be worth living. When the dreams and
aspirations of the youthful world have lost their old
significance and their strong power to raise men's
lives, it will be well for that community which has
organised in time a following of and a reverence for
an ideal Good, which may serve to lift the national

mind above the level of sensuality, and to insure a belief in the hopefulness and worth of life. The faith in Science can fill this place—the progress of Science is an ideal Good, sufficient to exert this great influence.

It is for this reason more than any other, as it seems to those who hold this faith, that the progress and diffusion of scientific research, its encouragement and reverential nurture, should be a chief business of the community, whether collectively or individually, at the present day.

APPENDIX

(A)

THE six years which have passed since this address was delivered have been remarkable for the extraordinary advances in our knowledge of the Bacteria and their relation to disease. In another paper (No. III.) I have given some account of M. Pasteur's discoveries in regard to rabies and hydrophobia. In the Pasteur Institute in Paris Dr. Roux and other investigators have greatly advanced our knowledge of the action of Bacteria in tuberculosis, in diphtheria, and other diseases. It now appears that the most hopeful method of combating the attacks of disease-producing Bacteria is not to attempt to introduce drugs into the infected patient which shall directly poison the Bacteria, but rather to endeavour to assist those natural processes in the body by which intrusive Bacteria are normally destroyed. Elias Metschnikoff has shown that the amœboid corpuscles of the blood and other tissues are special agents or scavengers for the destruction of such intrusive bodies as Bacteria. His theory is that by accustoming these corpuscles, which he calls "phagocytes," to tolerate a weak form of the poison produced by pathogenic Bacteria, we

" educate " them, so that they are able subsequently
to resist and eventually to attack and destroy the
same pathogenic Bacterium when present in a stronger
and deadly form.   This process of education is suggested as the rationale of preventive inoculation, the
" vaccination " with a modified cultivated growth of
poisonous Bacteria acting so as to educate the phagocytes, and enabling them subsequently to swallow and
digest the stronger infective growth.   By the training
afforded by inoculations with modified virus the phagocytes are, there is reason to believe, brought into a
condition of immunity similar to that which we know
the human body can attain in regard to other poisons,
such as laudanum, tobacco, and arsenic.   It was thus,
according to tradition, that Mithridates, King of
Pontus, failed at the end of his career to kill himself
with poison, having previously in his researches on
poisonous agents so saturated himself with them, and
habituated himself to their action, that none were
deadly for him any more.   We may then speak of
this training in tolerance of poison as " mithridatism,"
and refer to the trained phagocytes of an inoculated
animal as mithridatised.   On the other hand, there
are facts with regard to some kinds of infective virus
and their appropriate " vaccins " which lead to the
supposition that immunity is conferred, not by the
action of weak doses of the " toxin " or actual poison
of the virus, but by the action of a distinct chemical
substance which exists side by side with the " toxin."
The appendix to the paper on Pasteur and Hydrophobia, p. 166, contains a brief account of this theory.

It is an interesting fact in this connection that
poisonous snakes are, as shown by Sir Joseph Fayrer,
"mithridatised" in regard to their own poison.   The
bite of a poisonous snake has no effect on another
poisonous snake of the same group.   A similar discovery has been made by Professor Bourne of Madras,
who examined scorpions in regard to this matter at
my request.   A scorpion cannot sting itself to death,
nor can it sting another scorpion.   The scorpion is
"immune" to scorpion poison.   The explanation in
the case of both snakes and scorpions is this—that
small quantities of the poison elaborated by the poison-
glands in these animals are continually absorbed by
the blood, and thus the animal is mithridatised.   It
is known that the elaborated secretion of other glands
passes continually in small quantities into the blood
in other animals examined by physiologists; thus
traces of the secretion of the salivary gland and of
the pancreas are detected in the blood of rabbits and
dogs.

(B)

THE statement made as to the preponderating
activity of German scientific laboratories remains true,
although France has largely increased her already
extensive provision for scientific research by State aid.
The Institut Pasteur in Paris is the most important
scientific institution which has been founded of late
years.   It owes its existence partly to public subscription and partly to Government support.   Besides

providing salaries and laboratories for investigators who are carrying on the line of research in regard to Bacteria commenced by Pasteur, the Institute provides the preventive inoculation against hydrophobia discovered by Pasteur to all comers free of charge. Two hundred and fifty persons have gone from Great Britain and Ireland to be treated for dog-bite by M. Pasteur in Paris. The British Government has not made the smallest attempt to safeguard the public health by providing for the maintenance of such a laboratory as the Institut Pasteur in the United Kingdom. On the contrary, it has placed vexatious and ill-considered restrictions upon the conduct of experimental inquiries by competent physiologists in this country. These restrictions, without any doubt, prevent the discovery of new remedies for deadly disease and suffering.

(C)

THE question of State aid to science discussed in the preceding address is one which, although of the very greatest importance, does not make much progress.

There is at present a persistent confusion in the public mind, and even in the minds of statesmen and leading publicists, between the question of State aid to education and that of State aid to the production of new knowledge. It is no doubt true that if the professorships in provincial colleges and in University and King's Colleges are endowed by the State or other benefactor, the holders of those posts will possibly be able to, and probably will, give a larger amount of time than heretofore to the making of new knowledge instead of to the drudgery of journalism, popular lecturing, and cram-book writing. On the other hand, inadequate endowment of these colleges, or the granting to them of a small annual sum for general purposes not specified, will not necessarily have any result of the kind. The recent grant by the Government to university colleges of sums ranging from £500 to £2000 a year cannot be regarded as an endowment of research. In most cases the extremely meagre sum thus provided will be absorbed in general expenses of building and management connected with teaching. In other cases it will be wasted (*i.e.* returned to the giver—the tax-payer) by applying it to the reduction of pupil's fees. It is doubtful whether any college will have the courage to apply it deliberately to the payment of a professor or his assistant, so as to procure the production of new knowledge, or what is sometimes called " original research."

Since the date of my Southport address, University College has received a munificent bequest from one of its former professors, the late Mr. Richard Quain. Endowments of £300 a year have been assigned from this fund to the professorships of Botany, Physics, and English. It is doubtful whether these stipends should be considered as designed to enable the professor to devote his time to original researches, or as part payment for his labours in teaching. It is unfortunately not customary to distinguish the two occupations of making and distributing knowledge. In any case it cannot be considered that the endowment assigned to professorships in London is excessive in amount, or such as to secure the whole time of a professor, apart from his lectures, for original research.

A considerable expenditure made by the Government in direct aid of scientific research remains unnoticed in the text. I allude to the expenses of the Challenger Expedition, and the publication of the results obtained from the study of the collections brought home. This sum may be variously estimated, according to the sale of the reports and the inclusion or exclusion of certain official expenses, but in any case it amounts to several thousand pounds.

A very gratifying indication of the recognition on the part of the Government of the propriety of subsidising from the public funds scientic inquiries likely to be of general public utility was seen in the granting by Parliament in 1884-85, on the recommendation of the Treasury, of a sum of £5000 and an annual sum of £500 for five years towards the building and maintenance of the Plymouth Laboratory of the Marine Biological Association, of which some account will be found in the appendix to article No. V.

III

PASTEUR AND HYDROPHOBIA

FROM THE *Nineteenth Century*, JULY 1886

THE public has very naturally and very rightly shown deep interest in the investigations into the nature and possible cure of hydrophobia now being conducted by the great French naturalist, Louis Pasteur. Those investigations not only have a special value on account of the terrible nature of the malady which there is good reason to believe will be brought within the range of curative treatment as a consequence of their prosecution, but also are of extreme interest to those engaged in the task of ascertaining the laws of natural phenomena, and to all who wish to understand the methods by which a great discoverer in science arrives at his results.

M. Pasteur is no ordinary man ; he is one of the rare individuals who must be described by the term "genius." Having commenced his scientific career and attained great distinction as a chemist, M. Pasteur was led by his study of the chemical process of fermentations to give his attention to the phenomena of disease in living bodies resembling fermentations. Owing to a singular and fortunate mental characteristic he has been able, not simply to pursue a rigid path of investigation dictated by the logical or natural connection of the phenomena investigated, but deliberately to select for inquiry matters of the most profound importance to the community, and to bring his inquiries to a successful practical issue in a large number of instances. Thus he has saved the silk-worm industry of France and Italy from destruction, he has taught the French wine-makers to quickly mature their wine, he has effected an enormous improvement and economy in the manufacture of beer, he has rescued the sheep and cattle of Europe from the fatal disease "anthrax," and it is probable—he would not himself assert that it is at present more than probable—that he has rendered hydrophobia a thing of the past. The discoveries made by this remarkable man would have rendered him, had he patented their application and disposed of them according to commercial principles, the richest man in the world. They represent a gain of some millions sterling annually to the community. It is right for those who desire that increased support for scientific investigation should be afforded by the Governments of civilised States to point with emphasis to the definite utility and pecuniary value of M. Pasteur's work, because it is only in rare instances that the discovery of new knowledge and the practical application of that knowledge go hand in hand. M. Pasteur has afforded several of these rare instances. They should enable the public and our statesmen to believe in the value of scientific investigation even when it is not immediately followed by practical commercial results. These discoveries should excite in the minds

of all those devoted to scientific research the profoundest gratitude towards M. Pasteur, since, by the
direct practical application which his genius has
enabled him to give to the results of his inquiries, he
has done more than any living man to enable the
unlearned to arrive at a conception of the possible
value of the vast mass of scientific results—items of
new knowledge—which must be continually gathered
by less gifted individuals and stored for the future
use of inventors and of those doubly-gifted men
who, like M. Pasteur, are at once discoverers and
inventors—discoverers of a scientific principle and
inventors of its application to human requirements.

M. Pasteur's first experiment in relation to hydrophobia was made in December 1880, when he inoculated two rabbits with the mucus from the mouth of
a child which had died of that disease. As his inquiries extended he found that it was necessary to
establish by means of experiment even the most
elementary facts with regard to the disease, for the
existing knowledge on the subject was extremely
small, and much of what passed for knowledge was
only ill-founded tradition.

So little was hydrophobia understood, and to so
small an extent had it been studied, previously to
M. Pasteur's investigations, that it was regarded by a
certain number of highly competent physicians and
physiologists (although this was not the general view)
as a condition of the nervous system brought about by
the infliction of a punctured inflammatory wound in

which the action of a specific virus or poison took
no part; it was, in fact, by some physicians regarded
as a variety of lock-jaw or *tetanus*.

The number of cases of hydrophobia reported in
England, France, Germany, and Austria has varied a
good deal each year since the time when statistics of
disease were instituted by the Governments of these
several countries; but its occurrence is sufficiently
frequent at certain periods to excite the greatest
anxiety and alarm. In England as many as thirty-
six persons died from the disease in 1866; in France
288 persons were its victims in 1858, and in Prussia
and Austria it has been more frequent than in
England.

The general belief, both among medical men and
veterinary surgeons, as well as the public, has been
that the condition known as hydrophobia in man does
not follow from any ordinary bite or injury, but
that in order to produce it the human subject must
be bitten by a dog, wolf, pig, or other animal which
is suffering from a well-marked disease known as
"rabies." What it is which starts "rabies" amongst
dogs is not known, and has not even been guessed at,
but the condition so named is communicated by
"rabid" or "mad" dogs to other dogs, to pigs, to
cattle, to deer, to horses, and indeed to all warm-
blooded animals—even birds. Any animal so infected
is capable by its bite of communicating the disease to
other healthy animals. Rabies in a dog is recognised
without difficulty by the skilled veterinarian. The
disease has two varieties, known as "dumb madness"

and "raving madness"; and it is held by veterinarians
to have two modes of origin—viz. spontaneous, and as
the result of infection from another rabid animal.   It
is quite permissible to doubt the spontaneous generation of rabies in any given case, although it must be
admitted that the disease had a beginning, and that it
is not improbable that whatever conditions favoured
its first origin are still in operation, and likely to
result in a renewed creation of the disease from time
to time.   The disease was well known in classical
antiquity, and is of world-wide distribution, occurring
both in the tropics and in the arctic regions, though
much commoner in temperate regions than in either
of the extremes of climate.   There are some striking
cases of certain well-peopled regions of the earth's
surface in which it is at present unknown ; no case
appears to be on record of its occurrence in Australia,
Tasmania, or New Zealand.   It is a mistake to suppose that the disease is commoner in very hot weather
than in cooler weather, or that great cold favours it.
Climate, in fact, appears to have nothing to do with
it, or rather, it should be said, is not shown to have
anything to do with it.

Professor Fleming, in his admirable treatise on
Rabies and Hydrophobia (London, 1872), says :—

It is a great and dangerous error to suppose that the disease
(in the dog) commences with signs of raging madness, and that
the earliest phase of the malady is ushered in with fury and
destruction.   The first perceptible or initial symptoms of rabies
in the dog are related to its habits.   A change is observed in
the animal's aspect, behaviour, and external characteristics.   The
habits of the creature are anomalous and strange.   It becomes

dull, gloomy, and taciturn; seeks to isolate itself, and chooses
solitude and obscurity — hiding in out-of-the-way places, or re-
tiring below chairs and other pieces of furniture; whereas in
health it may have been lively, good-natured, and sociable. But
in its retirement it cannot rest; it is uneasy and fidgety, and
betrays an unmistakable state of *malaise;* no sooner has it lain
down and gathered itself together in the usual fashion of a dog
reposing than all at once it jumps up in an agitated manner,
walks hither and thither several times, again lies down, and
assumes a sleeping attitude, but has only maintained it for a few
minutes when it is once more moving about, "seeking rest but
finding none." Then it retires to its obscure corner — to the
deepest recess it can find—and huddles itself up in a heap, with
its head concealed beneath its chest and its forepaws. This
state of continual agitation and inquietude is in striking contrast
with its ordinary habits, and should, therefore, attract the attention of mindful people. Not unfrequently there are a few
moments when the creature appears more lively than usual,
and displays an extraordinary amount of affection. Sometimes
in pet dogs there is evinced a disposition to gather up small
objects, such as straws, threads, bits of wood, etc., which are
industriously picked up and carried away. A tendency to lick
anything cold, as iron, stones, etc., is also observed in many
instances. At this period no propensity to bite is observed;
the animal is docile with its master, and obeys his voice, though
not so readily as before, nor with the same pleased countenance.
If it shakes its tail the act is more slowly performed than usual,
and there is something strange in the expression of the face;
the voice of its master can scarcely change it for a few seconds
from a sullen gloominess to its ordinary animated aspect; and
when no longer influenced by the familiar talk or presence it
returns to its sad thoughts, for—as has been well and truth-
fully said by Bouley—"the dog thinks and has its own ideas,
which for dogs' ideas are, from its point of view, very good
ideas when it is well."

The animal's movements, attitudes, and gestures now seem
to indicate that it is haunted by and sees phantoms; it snaps at
nothing and barks as if attacked by real enemies. Its appearance is altered; it has a gloomy and somewhat ferocious aspect.

In this condition, however, it is not aggressive so far as mankind is concerned, but is as docile and obedient to its master as before. It may even appear to be more affectionate towards those it knows, and this it manifests by the greater desire to lick their hands and faces.

This affection, which is always so marked and so enduring in the dog, dominates it so strongly in rabies that it will not injure those it loves, not even in a paroxysm of madness; and even when its ferocious instincts are beginning to be manifested, and to gain the supremacy over it, it will yet yield obedience to those to whom it has been accustomed.

The mad dog has not a dread of water, but, on the contrary, will greedily swallow it. As long as it can drink it will satisfy its ever-ardent thirst; even when the spasms in its throat prevent it swallowing, it will nevertheless plunge its face deeply into the water and appear to gulp at it. The dog is, therefore, not hydrophobic, and hydrophobia is not a sign of madness in this animal.

It does not generally refuse food in the early period of the disease, but sometimes eats with more voracity than usual.

When the desire to bite, which is one of the essential characters of rabies at a certain stage, begins to manifest itself, the animal at first attacks inert bodies—gnawing wood, leather, its chain, carpets, straw, hair, coals, earth, the excrement of other animals or even its own, and accumulates in the stomach the remains of all the substances it has been tearing with its teeth.

An abundance of saliva is not a constant symptom in rabies in the dog. Sometimes its mouth is humid, and sometimes it is dry. Before a fit of madness the secretion of saliva is normal; during this period it may be increased, but towards the end of the malady it is usually decreased.

The animal often expresses a sensation of inconvenience or pain during the spasm in its throat by using its paws on the side of its mouth, like a dog which has a bone lodged there.

In "dumb madness" the lower jaw is paralysed and drops, leaving the mouth open and dry, and its lining membrane exhibiting a reddish-brown hue; the tongue is frequently brown or blue coloured, one or both eyes squint, and the creature is ordinarily helpless and not aggressive.

In some instances the rabid dog vomits a chocolate or blood-coloured fluid.

The voice is always changed in tone, and the animal howls or barks in quite a different fashion to what it did in health. The sound is husky and jerking. In "dumb madness" this very important symptom is absent.

The sensibility of the rabid dog is greatly blunted when it is struck, burned, or wounded ; it emits no cry of pain or sign as when it suffers or is afraid in health. It will even sometimes wound itself severely with its teeth, and without attempting to hurt any person it knows.

The mad dog is always very much enraged at the sight of an animal of its own species. Even when the malady might be considered as yet in a latent condition, as soon as it sees another dog it shows this strange antipathy and appears desirous of attacking it. This is a most important indication.

It often flees from home when the ferocious instincts commence to gain an ascendency, and after one, or two, or three days' wanderings, during which it has tried to gratify its mad fancies on all the living creatures it has encountered, it often returns to its master to die. At other times it escapes in the night, and after doing as much damage as its violence prompts it to, it will return again towards morning. The distances a mad dog will travel, even in a short period, are sometimes very great.

The furious period of rabies is characterised by an expression of ferocity in the animal's physiognomy, and by the desire to bite whenever an opportunity offers. It always prefers to attack another dog, though other animals are also victims.

The paroxysms of fury are succeeded by periods of comparative calm, during which the appearance of the creature is liable to mislead the uninitiated as to the nature of the malady.

The mad dog usually attacks other creatures rather than man when at liberty. When exhausted by the paroxysms and contentions it has experienced, it runs in an unsteady manner, its tail pendent and head inclined towards the ground, its eyes wandering and frequently squinting, and its mouth open, with the bluish-coloured tongue, soiled with dust, protruding.

In this condition it has no longer the violent aggressive

tendencies of the previous stage, though it will yet bite every one
—man or beast—that it can reach with its teeth, especially if
irritated.

The mad dog that is not killed perishes from paralysis and
asphyxia.   To the last moment the terrible desire to bite is
predominant, even when the poor creature is so prostrated as to
appear to be transformed into an inert mass.

Such is the pathetic account of the features of this
terrible malady as seen in man's faithful companion.
Let us now for a moment look at the symptoms and
course of the disease as exhibited in man—where it
produces a condition so terrible and heart-rending to
the onlooker that it becomes a matter of astonishment
that mankind has ever ventured to incur the risk
of acquiring this disease by voluntarily associating
with the dog, and a matter of the most urgent desire
that some great deliverer should arise and show us how
to remove this awful thing from our midst.

In both the dog and man the disease is traced
to the infliction of a bite or scratch at a more or less
distant period by an animal already suffering from
rabies.   The length of time which may elapse between
the bite and the first symptoms of " rabies " in the dog
or of " hydrophobia," as it is termed, when developed
in man, varies.   Briefly, it may be stated that the
interval in the dog varies from seven to one hundred
and fifty days, and is as often a longer as a shorter
period.   In man, on the other hand, two-thirds of the
cases observed develop within five weeks of the
infliction of the infecting bite ; hydrophobia *may*
show itself as early as the eighth day after the
infection ; it is very rare indeed, though not unknown,

that this period of incubation is extended to a whole year. The reputed cases of an "incubation period" of two, five, or even ten years may be dismissed as altogether improbable and unsupported by evidence. The uncertainty which this well-known variation in the incubation period produces is one of the many distressing features of the disease in relation to man, for often the greatest mental torture is experienced during this delay by persons who after all have not been actually infected.

In many respects (says Professor Fleming) there is a striking similarity in the symptoms manifested in the hydrophobic patient and the rabid dog, while in others there is a wide dissimilarity. These resemblances and differences we will note as we proceed to briefly sketch the phenomena of the disease in our own species.

The period of incubation or latency has been already alluded to, and it has also been mentioned that not unfrequently in man and the dog the earliest indication of approaching indisposition is a sense of pain in or near the seat of the wound, extending towards the body, should the injury have been inflicted on the limbs. If not acute pain there is some unusual sensation, such as aching, tingling, burning, coldness, numbness, or stiffness in the cicatrix ; which usually, in these circumstances, becomes of a red or lurid colour, sometimes opens up, and if yet unhealed assumes an unhealthy appearance, discharging a thin ichorous fluid instead of pus. In the dog, as we have observed, the peculiar sensation in the seat of the inoculation has at times caused the animal to gnaw the part most severely.

With these local symptoms some general nervous disturbance is generally experienced. The patient becomes dejected, morose, irritable, and restless ; he either does not suspect his complaint, or, if he remembers having been bitten, carefully avoids mention-ing the circumstance, and searches for amusement away from home, or prefers solitude ; bright and sudden light is disagreeable to him ; his sleep is troubled, and he often starts up ; pains are experienced in various parts of the body; and signs of digestive disorder are not unfrequent.   After the continuance of one or more of these preliminary, or rather premonitory, symptoms for a period varying from a few hours to five or six days, and, though very rarely, without all or even many of them being observed, the patient becomes sensible of a stiffness or tightness about the throat, rigours supervene, and in attempting to swallow he experiences some difficulty, especially with liquids.   This may be considered as really the commencement of the attack in man.

The difficulty in swallowing rapidly increases, and it is not long before the act becomes impossible, unless it is attempted with determination; though even then it excites the most painful spasms in the back of the throat, with other indescribable sensations, all of which appeal to the patient, and cause him to dread the very thought of liquids.   Singular nervous paroxysms or tremblings become manifest, and sensations of stricture or oppression are felt about the throat and chest.   The breathing is painful and embarrassed, and interrupted with frequent sighs or a peculiar kind of sobbing movement; and there is a sense of impending suffocation and of necessity for fresh air.   Indeed, the most marked symptoms consist in a horribly violent convulsion or spasm of the muscles of the larynx and gullet, by which swallowing is prevented, and at the same time the entrance of air to the windpipe is greatly retarded.   Shuddering tremors, sometimes almost amounting to general convulsions, run through the whole frame; and a fearful expression of anxiety, terror, or despair is depicted on the countenance.

The paroxysms are brought on by the slightest causes, and are frequently associated with an attempt to swallow liquids, or with the recollection of the sufferings experienced in former attempts.   Hence anything which suggests the idea of drinking to the patient will throw him into the most painful agitation and convulsive spasms. . . . This is particularly observed when the patient carries water to his lips; then he is seized with the terrors characteristic of the disease, and with those convulsions of the face and the whole of the body which make so deep an impression on the bystanders.   He is perfectly rational, feels thirsty, tries to drink, but the liquid has no sooner touched his lips than he draws back in terror, and sometimes exclaims that he cannot drink ; his face expresses pain, his eyes are fixed, and his features contracted ; his limbs shake and body trembles. The paroxysm lasts a few seconds, and then he gradually becomes tranquil ; but the least touch, nay, mere vibration of the air, is enough to bring on a fresh attack—so acute is the sensibility of the skin in some instances. . . . A special difference between rabies and hydrophobia is the frequent dread of water in the latter, as well as the hyperæsthesia of the skin and exaltation of the other senses. . . . Another characteristic feature of the disease in man is a copious secretion of viscid, tenacious mucus in the fauces, the "hydrophobic slaver" ; this the patient spits out with a sort of vehemence and rapidity upon everything around him, as if the idea of swallowing occasioned by the liquid induced this eager expulsion of it, lest a drop might pass down the throat. This to a bystander is sometimes one of the most striking phenomena of the case. . . . The mind is sometimes calm and collected in the intervals between the paroxysms, and consciousness is generally retained ; but in most cases there is more or less irregularity, incessant talking, excitement, and occasionally fits approaching to insanity come on. The mental aberration is often exhibited in groundless suspicion or apprehension of something extraneous, which is expressed on the face and in the manner of the patient. In comparatively rare instances he gives way to a wild fury, like that of a dog in one of its fits of rabies ; he roars, howls, curses, strikes at persons near him, rends or breaks everything within his reach, bites others or himself, till, at length exhausted, he sinks into a gloomy, listless dejection, from which another paroxysm rouses him. . . . Paralytic symptoms manifest themselves before death in a few instances, as in the dog. . . . Remissions of the symptoms sometimes occur in the course of the complaint, during which the patient can drink, though with some difficulty, and take food. Towards the close such a remission is not uncommon, with an almost complete absence of the painful symptoms ; so that the patient and the physician begin to entertain some hope. But if the pulse is now felt it is found to be extremely feeble, and sometimes almost, if not quite, imperceptible. During this apparent relaxation of the disease the patient occasionally falls into a sleep, from which he only awakes to die.

Death results from spasm of the respiratory muscles, the patient dying asphyxiated. The desire to bite is rare. The disease invariably, as in the dog and other animals, terminates fatally, and usually between the second and fifth day after the symptoms have been first observed, though it sometimes runs on to the ninth day.

It is held by veterinaries that "rabies" in a dog is invariably fatal, and one test of the presence of the disease is a fatal termination to the symptoms. Inasmuch as it is very usual to kill dogs suspected of rabies without waiting to actually prove that they suffer from this disease, and further, inasmuch as dogs not suffering from rabies are nevertheless frequently savage or snappish and bite human beings, thus leading to the assumption that the person bitten has incurred the risk of developing hydrophobia, there is necessarily a complete absence of trustworthy statistical information as to (1) the actual number of dogs annually affected with rabies in any given country, and (2) as to the number of persons effectively bitten by really rabid dogs, who acquire hydrophobia as a consequence. The dogs are killed before it is proved that they suffer from rabies, and the human beings bitten are treated by caustics and excision of injured surfaces before it is proved that they really are in danger of developing hydrophobia, and it is not known in case of escape whether the danger was ever really incurred. The extreme anxiety to avoid the awful consequences not unfrequently following the bite of a rabid dog has produced a course of action which, whilst it is undoubtedly accompanied by the destruction of many innocent dogs, and by the infliction of acute pain and mental anguish upon human beings, who, could they know the truth, have no cause for alarm, has also at the same time necessarily prevented the acquisition of accurate knowledge with regard to the disease in important respects, especially as to the conditions of its communication from dog to man.   Accordingly, we find great uncertainty as to the conclusions which are to be drawn from statistics in regard to the effect on human beings of the bites of dogs suffering from rabies.   According to the lowest estimate where care has been taken to exclude cases in which there is insufficient reason for supposing the offending dog to have suffered from rabies, of every *six* persons bitten, *one* dies—that is to say, *one* develops hydro-phobia; for recovery after the development of the hitherto recognised symptoms of hydrophobia is un-known.   This is a mortality of 16·66 per cent; other estimates range from 15 to 25 per cent.   The large proportion of escapes as compared with deaths is attributed to the wounds inflicted not having been sufficiently deep to introduce the poison into the system, also to timely surgical treatment having the same effect, and to the fact that the dog, in spite of probabilities to the contrary, may in a certain proportion of cases have been wrongly suspected of suffering from " rabies."

At the same time there is no doubt that animals (and hence presumably man) are sometimes endowed with an immunity from rabies.   This has been proved

experimentally by repeatedly inoculating a dog with
the saliva of rabid dogs which proved fatal to other
individuals which were experimented upon at the
same time, whilst the particular dog in question
always proved refractory or non-liable to the disease.
No estimate has been at present formed of the proportion of dogs which are thus free from liability
to the disease, but it must be very small, perhaps not
1 per cent.   On the other hand, it is undeniable that
there is a high probability that such immunity exists
among human beings, and it is possible that the
proportion of individuals liable to the infection as
compared with those "immune," "refractory," or "non-
liable" is less amongst human beings than among
dogs.   Such a constitutional immunity may, therefore,
possibly explain to a certain extent the fact that out
of 100 cases of dog-bite, the dog being supposed,
but not demonstrated, to be rabid, only 16 acquire
hydrophobia.

The result of M. Pasteur's experimental study of
rabies and hydrophobia has been so far to place several
matters of practical importance, which were previously
liable to be dealt with by vague guesses and general
impression, in the position of facts capable of accurate
experimental determination ; and secondly, to introduce a method of treating animals and men infected
with the poison of rabies in a way which, there is
strong evidence to show, will arrest or altogether prevent the development of the disease.

Owing to the eagerness of newspaper correspond-
ents, and the peculiar circumstances of the investiga-

tion which is still actually in progress, M. Pasteur's work has been not quite fairly represented to the public, and various astonishing criticisms and expressions of individual opinion have been indulged in, with regard to what M. Pasteur is doing, by persons who, however gifted, have no adequate comprehension of the task which the great experimenter has set before himself.

It must be distinctly remembered, on the one hand, that the results which M. Pasteur has himself published, and for which he has made himself responsible, have been obtained by accurate and demonstrative experiments upon animals ; they are results which can be repeated and verified.   On the other hand, M. Pasteur has now advanced into a much more difficult field— namely, the application of his experimentally ascertained results to the treatment of human beings.   He is actually in course of carrying out his inquiries in regard to the efficacy of his treatment, and it is probable that at no distant date he will himself give us a detailed account of the conclusions to which these inquiries lead.   But he has not yet formulated any such conclusion.

We cannot and have not the remotest desire to experiment upon human beings, as in the more enlightened parts of Europe we are permitted, for good purposes, to experiment upon dogs.   It is not possible to exactly arrange experimentally the conditions of a human being who is to be the subject of inquiry in regard to hydrophobia.   You cannot make sure by the inoculation in the most effective way of a dozen

healthy men that they have started on the path
leading to hydrophobia, and then treat six by a
remedial process, and leave six without such treatment, in order to see whether the remedial process
has an effect or not. This is the kind of difficulty which is met with in all attempts to take a
step forward in medical treatment. Nevertheless,
although such definite experimental arrangement of
the subject of inquiry is not possible where human
beings are concerned, there is another method—
extremely laborious, and less decisive in the results
which it affords—by which a more or less probable
conclusion may be arrived at in regard to the effect
of treatment of diseased human beings. This method
consists in bringing together for experimental treatment a very large number—some thousands—of cases
in which the disease under investigation has, independently of the experimenter, been acquired, or is supposed
to have been acquired, and then to compare the proportion of cases of recovery obtained under the new
treatment with the proportion of recoveries in cases
not subjected to this treatment.

Hydrophobia presents peculiar difficulties in the
application of this method, and the treatment which
M. Pasteur is now testing is also one which in its
essence renders the statistical method difficult of
application. M. Pasteur's treatment has to be applied
before the definite symptoms of hydrophobia have
developed in the patient. Accordingly, there is no
certain indication in the patient himself that he has
really been infected by the virus of rabies ; the infer-

ence that he has been so infected is based on the
knowledge of the condition of the dog that bit the
patient, and on the extent of the injury inflicted ; but
the knowledge of the actual state of the dog which
inflicted the bite upon a person who, therefore, has
reason to fear an attack of hydrophobia is often wanting.   It is often merely "feared" or "supposed" that
the dog was rabid, and has not been actually proved
that such was the case.   In many cases the only proof
that the dog really was rabid would be found in the
development of hydrophobia in the man bitten by the
dog, the dog itself having been destroyed.   This, too,
would be the only definite proof possible that the
patient had received a sufficiently profound wound to
carry the poison into the system, or, again, that the
patient is not naturally "immune" or "refractory" to
the poison.   Accordingly, it has been necessary for
M. Pasteur to test his treatment upon a very large
number of cases, so as to obtain a statistical result
which may be compared with the general statistics of
the effects following the bite of reputed rabid dogs.
Also, it is possible out of a large number of cases for
M. Pasteur to select, without any other determining
motive, those cases in which the dog which inflicted
the bite was actually proved to be suffering from
rabies, either by the result of its bite on other individuals, or by experiments made by inoculating other
animals from it after its death.   Such a selection of
his cases has, it is stated, already been made by M.
Pasteur.   We have yet to await from M. Pasteur's
own hand a critical account of the results obtained in

the wholesale treatment of patients by him in Paris. Until he has himself published that account, we ought to be very careful about coming to an absolute conclusion either for or against the efficacy of his treatment *in regard to men.*

On the other hand, the fundamental results of his study of rabies and hydrophobia stand in no such position, but are sharp, experimental demonstrations, which he has publicly announced before the scientific world, and has verified in the most important instance before a commission appointed by the Government.

Let us note some of these results.[1] They have been obtained by experimentally inoculating dogs, rabbits, guinea-pigs, and monkeys. The experiments have been performed by M. Pasteur himself and his experienced and highly - skilled assistants, MM. Chamberland and Roux. Precautions which a thorough knowledge of the subject suggested have been taken. Thus, for instance, in his very first experiments, M. Pasteur cleared the ground considerably by distinguishing a kind of blood-poisoning, due to the presence of a certain bacterium in human saliva, which is liable to be introduced with the saliva of a hydrophobic patient when this is made use of for the purpose of setting up rabies experimentally in a rabbit, and is also present in normal saliva. Not feeling sure that some rabbits thus treated had really died from rabies, and suspecting that they might have died from a

---

[1] I am indebted to an excellent report by my friend Dr. Vignal, of the Collège de France, published in the *British Medical Journal,* for the chief facts relative to M. Pasteur's published results.

blood-poisoning due to another virus present in the hydrophobic saliva, M. Pasteur tested his rabbits by inoculating dogs with the saliva and blood of the rabbits. The dogs did not develop rabies, and thus M. Pasteur was able to establish the conclusion confirmed by other observations—that the disease produced in this instance by the inoculation of the rabbits with saliva was not rabies. This is merely an example of the careful method by which it is M. Pasteur's habit to correct and solidly build up his conclusions.

The first result of great practical moment established by M. Pasteur is that not only, as shown by previous experimenters, can rabies be communicated from animal to animal by the introduction of the saliva of a rabid animal into the loose tissue beneath the skin of a healthy animal, or by injection of the same into the veins of a healthy animal, but that the "virus," or poison, which carries the disease resides in its most active form in the nervous tissue of a rabid animal, and that the most certain method of communicating rabies from one animal to another is to introduce a piece of the spinal cord or of a large nerve of a rabid animal on to the surface of the brain of a healthy animal, the operation of exposing the brain being performed with the most careful antiseptic methods, so as to prevent blood-poisoning.

In this way Pasteur found that he could avoid the complications which sometimes result from the presence of undesired poisonous matters—not related to rabies—in the saliva of rabid animals.

This discovery is the starting-point of all Pasteur's further work. It enabled him to experiment with sufficient certainty as to results. It has rendered it possible for him to determine whether a dog is really affected with rabies or not, by killing it and inoculating the brain of a second dog with the spinal cord of the dead dog, and similarly to determine whether a human being has really died of hydrophobia (*rabies hominis*) or not. It has also enabled him to propagate with certainty the disease from rabbit to rabbit through ninety successive individuals—extending over a period of three years—and to experiment on the result of varying the quantity of virus introduced as well as on the result of passing the virus from one species of animal to another, and back again to the first species (*e.g.* rabbit as the first and monkey as the second species). Before Pasteur's time Rossi, confirmed by Hertwig, had used nerve-tissue for inoculation with less definite results. Pasteur has the merit of establishing this method as the really efficient one in experimenting on the transmission of rabies.

Using the nerve-tissue, Pasteur has determined by several experiments that when a large quantity of virus (that is to say, of the medulla oblongata of a rabid rabbit pounded up in a perfectly neutral or sterilised broth) is injected into the veins of a dog, the incubation period is seven or eight days ; by using a smaller quantity he obtained an incubation period of twenty days, and by using a yet smaller quantity one of thirty-eight days. It is very important to note that by using a still smaller dose Pasteur found that the dog so treated escaped the effect of the poison altogether.

A very interesting and important result is that in the cases in which the largest amount of poison was used, and the quickest development of the disease followed, the form which the disease took was that of paralytic or "dumb rabies," in which the animal neither barks nor bites; whilst with the smaller dose of poison and longer incubation period "furious rabies" was developed.   Moreover, by directly inoculating on the surface of the brain and spinal cord, Pasteur has been led to the conclusion that the nature of the attack can be varied by the part of the central nervous system which is selected as the seat of inoculation.

Certain theories which have been held as to the mode in which inoculation with the attenuated virus of such diseases as small-pox and anthrax acts, so as to protect an animal from the effect of subsequent exposure to the full strength of the poison, might lead us to expect that the dogs which were inoculated by M. Pasteur with a quantity of rabid virus just small enough to fail in producing the symptoms of rabies would be "protected" by that treatment from the injurious effects of subsequent inoculation with a full dose.   This, however, Pasteur found was *not* the case. Such dogs, when subsequently inoculated with a full dose, developed rabies in the usual way.

When the virus of rabies is introduced from a dog into a rabbit, and is cultivated through a series of rabbits by inoculating the brain with a piece of the

spinal cord of a rabid animal, Pasteur has found that
the virulence of the poison is increased. The incubation period becomes shorter, being at first about fifteen
days. After being transmitted from rabbit to rabbit
through a series of twenty-five individuals, the period
of incubation becomes reduced to eight days, and the
virulence of the poison is proportionately increased.
After a further transmission through twenty-five
individuals, the incubation period is reduced to seven
days, and after forty more transmissions Pasteur finds
an indication of a further shortening of the incubation
period, and a proportionate increase of virulence in the
spinal cord of the rabbit extracted after death and
used for inoculating other animals. Thus Pasteur
found it possible to have at his disposal simultaneously
rabid virus of different degrees of activity.

It is curious that Pasteur found, on the other hand,
that the virus from a rabid dog, when transmitted
from individual to individual through a series of
monkeys, gradually lost its activity, so that after passing through twenty (?) monkeys it became incapable
of producing rabies in dogs. Thus a portion of the
spinal cord of such a monkey, itself dead of rabies,
when pounded in broth and injected beneath the skin
of a dog, failed to produce rabies, and even when applied to the dog's brain after trephining failed to produce rabies.

Pasteur makes the very important statement that
the dogs thus treated with the virus which had been
weakened by cultivation in monkeys, although they
did not develop any symptoms of rabies, *were ren-*

dered refractory to subsequent inoculations with strong virus—that is, were "protected."

Thus we note a contrast between the effect obtained by inoculating an animal with a virus weakened by cultivation, and those resulting from using a minute quantity of the virus. The latter proceeding does not result in protection, but the former does.

The fresh spinal cord of an animal that has died of rabies is apparently full of the rabid virus, and it will, if kept so as to prohibit putrefaction, retain for some days its rabies-producing property. Nevertheless it gradually, without any putrefactive change, loses, according to Pasteur's observations, its virulence, which finally disappears altogether. So that it is possible to obtain cord of a very low degree of virulence, and all intermediate stages leading up to the most active, by the simple process of suspending a series of cords at definite intervals of time in glass jars containing dry air.

There are thus two ways of bringing the virus of rabies taken from a dog into a condition of diminished activity—the one by cultivation in monkeys or some other animal, the other by exposing the spinal cord to dry air whilst preventing it from putrefying.

It was found by Pasteur that dogs inoculated with the virus weakened by cultivation in monkeys were protected from the effects of subsequent inoculation with strong virus. Hence he proceeded to experiment in the direction so indicated. He inoculated dogs with a very weak virus taken from a rabbit—that is, a virus having a long incubation period—and at the same time he inoculated also a rabbit. When the second rabbit went mad and died, the dogs were again inoculated from it, and a third rabbit was also inoculated from it. When this rabbit died the process was repeated with the dogs and with a fourth rabbit, and so on until the virus had become (as above stated to be the case) greatly increased in activity, its incubation period being reduced to eight days. The dogs were not rendered rabid by the first inoculations ; they certainly would have been by the last, had they not undergone the earlier. The harmless virus rendered the dogs insusceptible to the rabies-producing quality of the second dose introduced, the second did the same for the third, the third for the fourth, and so on until the dogs were able to withstand the strongest virus.

It would seem that this method of using a graduated series of poisons was not intentional on Pasteur's part at first, but merely arose from the convenience of the arrangement, since the effect of the previous inoculation could be tested and a new inoculation to act as a preventive could be made at one and the same time. Nevertheless, Pasteur has retained for reasons, which it is possible to imagine but have not been given as yet by him, this method of repeated doses of gradually increasing strength in his subsequent treatment.

In 1884 a Commission was appointed at M. Pasteur's request by the Minister of Public Instruction to examine the results so far obtained by him in regard to a treatment by which dogs could be rendered refractory to rabies. The Commission comprised some of the ablest physiologists in France ; it consisted of

MM. Béclard, Paul Bert, Bouley (the celebrated veterinarian), Tisserand, Villemin, and Vulpian. Their report contained the following statement:—

The results observed by the Commission may be thus summarised. Nineteen control dogs (*i.e.* ordinary dogs not treated by Pasteur) were experimented on. Among six of these bitten by mad dogs, three were seized with rabies. There were six cases of rabies among eight of them subjected to venous inoculations, and five cases of rabies among five which were inoculated by trephining on the brain. The twenty-three dogs treated (by Pasteur) and then tested all escaped rabies.[1]

Subsequently to the experiments witnessed by the Commission M. Pasteur carried out experiments in which, instead of using virus of increasing strength taken from living rabbits, he made use of the fact discovered by him that the spinal cord of a rabid animal when preserved in dry air retains its virulent property for several days, whilst the intensity of the virulence gradually diminishes. Pasteur used for this purpose cords of rabbits affected with rabies of great virulence, determined by a long series of transmissions, and having only an eight days' incubation period. He injected a dog on the first day with a cord which, when fresh, was highly virulent, but had been kept for ten days, and hence was incapable of starting rabies in the dog; on the second day he used a cord kept for nine days, on the third day a cord kept for eight

---

[1] I have ascertained that of these twenty-three dogs some had been already treated by Pasteur before the appointment of the Commission, and a minority were treated by him for the first time in the presence of the Commission. Ten of these dogs are still in M. Pasteur's hands, and have been inoculated three times on the surface of the brain with rabid virus; not one has developed rabies.—*July* 1887.

days, and so on until on the tenth day a cord kept
for only one day was used. This was found to cause
rabies in a dog not previously treated, and yet had
no such effect on the dog subjected to the previous
series of inoculations. The dog had been rendered
refractory to rabies. In this way M. Pasteur states
that he rendered fifty dogs of all ages and races refractory to (or "protected against") rabies *without one
failure*. Virus was inoculated under the skin and
even on the surface of the brain after trephining, and
rabies was not contracted in a single case.

Why M. Pasteur makes use of a gradually increasing strength of virus, or how he supposes this treatment to act so as to give the remarkable result of
protection, he has not explained. The experimenter
very probably has his own theory on the subject,
which guides him in his work ; but whilst he is still
experimenting and observing he does not commit himself to an explanation of the results obtained. We
may look in the future for a full consideration of the
subject and a definite statement of the evidence at his
hands. Meanwhile, it must be remembered that the
notes published by M. Pasteur are, as it were, bulle-
tins from the field of battle, briefly announcing failures
and successes, and are not to be regarded as a history
of the campaign or a statement of its scheme and final
result.

Having arrived at this point in his experimental
results, M. Pasteur was prepared to venture on to the
far more delicate ground of treatment of human beings
who had incurred the risk of hydrophobia.

The period of incubation of hydrophobia being usually four or five weeks, it seemed to M. Pasteur not impossible that he might succeed by the method which he had carried out in dogs in rapidly producing in human subjects a state of refractoriness to the poison of rabies by using a virus of rapid activity, and so, as it were, overtake the more slowly acting virus injected into the system by the bite of a mad dog.

Whatever may have been his theoretical conceptions, M. Pasteur determined to have recourse to the one great and fertile source of new knowledge—experiment.

It is known that inoculation with vaccine virus during the latent period of small-pox has an effect in modifying the disease in a favourable direction, and so in any case it was to be expected that the inoculation of individuals during the latent period of hydrophobia might produce favourable results. M. Pasteur had every reason to believe that, at any rate, the inoculation which he proposed would not have injurious results. He could proceed to the trial with a clear conscience, feeling sure that he was in any case giving the bitten person a better chance of recovery than he would have if left untreated.

The first human being treated by Pasteur was the child Joseph Meister, who was sent from Alsace by Dr. Weber and arrived in M. Pasteur's laboratory on the 6th of July 1885. This child had been bitten a few days previously, in fourteen different places, by a mad dog, on the hands, legs, and thighs. MM. Vulpian and Grancher, two eminent physicians, considered Meister to be almost certain to die of hydrophobia. M. Pasteur determined to treat the child by the method of daily injection of the virus of a series of rabbits' spinal cords, beginning with one kept so long as to be ineffective in the production of rabies even in rabbits, and ending with one so virulent as to produce rabies in a large dog in eight days.

On the 6th of July 1885, M. Pasteur inoculated Joseph Meister, under the skin, with a Pravaz's syringe half full of sterilised broth (this is used merely as a diluent), mixed with a fragment of rabid spinal cord taken from a rabbit which had died on the 21st of June. The cord had since that date been kept in a jar containing dry air—that is, fifteen days. On the following days, Meister was inoculated with spinal cord from rabid rabbits kept for a less period. On the 7th of July, in the morning with cord of fourteen days; in the evening with cord of twelve days; on the 8th of July, in the morning with cord of eleven days, in the evening with cord of nine days; on the 9th of July, with cord of eight days; on the 10th of July, with cord of seven days; on the 11th of July, with cord of six days; on the 12th of July, with cord of five days; on the 13th of July, with cord of four days; on the 14th of July, with cord of three days; on the 15th of July, with cord of two days; on the 16th of July, with cord of one day. The fluid used for the last inoculation was of a very virulent character. It was tested and found to produce rabies in rabbits with an incubation period of seven days; and in a normal healthy dog it produced rabies with an incubation period of ten days.

It is now twelve months since Joseph Meister was bitten by the mad dog, and he is in perfect health. Even if we set aside the original infection from the mad dog, we have the immensely important fact that he has been subjected to the inoculation of strong rabid virus by M. Pasteur and has proved entirely insusceptible to any injurious effects, such as it could and did produce in a powerful dog.

M. Pasteur now proceeded, immediately after Meister's case, to apply his method to as many persons as possible who had reason to believe that they had been infected by the virus of a mad dog or other rabid animal. It must be remembered that Pasteur does not attempt to treat a case in which hydrophobia has actually made its appearance, and that he would desire to begin his treatment as soon after the infection or bite as possible; the later the date to which the treatment is deferred, the less is the chance—naturally enough—of its proving effective. He now omits the first three inoculations of weakest quality used in the case of Joseph Meister, and makes only ten inoculations (beneath the skin on the abdomen), one every day for ten days, the strength of the virus being increased as above explained. Probably Pasteur is varying and improving his method in regard to certain details. He himself has made no statement of a conclusive nature during the year. He is observing and collecting his facts. But Dr. Grancher, who is at present Pasteur's chief assistant in carrying on the inoculations of human patients, has recently published a rough analysis of the cases treated.

It appears that between the 6th of July 1885 and the 10th of June 1886 the number of patients treated by Pasteur's method was 1335. In order to eliminate cases of which the final issue is uncertain, Dr. Grancher omits those treated subsequently to the 22d of April 1886. Of the cases treated within the period thus defined, there were ninety-six in which the patients had been bitten by dogs which were absolutely demonstrated to be suffering from rabies. This demonstration was afforded either by the fact that other animals bitten by them became rabid or by an experiment in which a portion of the dog's brain being placed in contact with the brain of a living rabbit was found to cause the death of that rabbit with indisputable symptoms of rabies. A second class of cases were those of persons who were bitten by dogs certified to be rabid by the veterinary practitioners of the locality in which the bite took place. Of these there were 644. Lastly, there were 232 cases in which the dog which had inflicted the bite had run off and not been seen again, leaving it entirely doubtful as to whether the dog had really been rabid or not.

For the purpose of judging of the efficacy of Pasteur's method the last group of cases should be put aside altogether. In the first two classes there are 740 cases. These we can compare with the most carefully formed conclusions as to the result of bites of rabid dogs when Pasteur's treatment has not been adopted. In the first part of this article it was stated that the inquiries of the most experienced veterinarians lead to the conclusion that 16 per cent of human beings who are bitten by dogs which are certified to be rabid by veterinary surgeons skilled in that disease, develop hydrophobia and die. This estimate is a low one; by some authorities 25 per cent has been regarded as nearer the true average. Taking the lower estimate, there should have died amongst Pasteur's 740 patients no less than 118.

What, then, is the difference resulting (so far as we can judge at present) from the application to these persons of Pasteur's method of treatment?

Instead of 118 deaths, there have been only 4, or a death-rate of one-half per cent instead of 16 per cent. In less than one year, it seems, Pasteur has directly saved 114 lives. When we remember what a death it is from which apparently he has saved those hundred and more men, women, and children, who can measure the gratitude which is due to him or the value of the studies which have led him to this result?

Nevertheless, let us be cautious. It is very natural that we should hasten to estimate the benefit which has been conferred on mankind by this discovery; on the other hand, the method of testing its value by comparative statistics is admittedly liable to error. Whilst the figures so far before us justify us in entertaining the most sanguine view, a longer series of cases will be needful, and *minute examination of each case*, before a final judgment can be pronounced. We have not before us at present the data for a more minute consideration of the separate cases. But one of the most hopeful features in M. Grancher's statement is that he records only one death out of the ninety-six persons who were bitten by dogs experimentally proved to be rabid—proved, that is, by the communication of rabies by the dogs to other animals.

Another extremely important series of cases is afforded by the forty-eight cases of wolf-bites treated by Pasteur's method. Owing to the fact that the rabid wolf attacks the throat and face of the man upon whom it rushes, the virus is not cleared from its teeth by their passage through clothing, as undoubtedly occurs in many cases of rabid dogs' bite. It is probable that this, together with the greater depth and extent of the wounds inflicted by wolves, accounts for the fact that whilst only 16 per cent of the persons bitten by rabid dogs die, as many as 66·5 per cent of the persons bitten by rabid wolves have hitherto succumbed. Pasteur has reduced this percentage in the forty-eight cases of wolf-bites treated by him to 14 ; seven of his cases died. But it is important to remember that some of these cases were treated a long while (three weeks or more) after the bite ; and also that the bites themselves, apart from the virus introduced into them, were of a very dangerous nature in some cases. On the other hand, it is equally true that we do not know, until some very much more complete record is placed before us than we have at present, how many cases of very slight injury, mere nips or scratches, may have been included among the forty-eight cases of wolf-bite.

Pasteur is still observing : he himself has not pronounced his method to be final, nor that its efficacy is actually so great as the figures above given would seem to indicate. Time will show ; meanwhile it is clear that the treatment is in itself harmless, and gives such reasonable hope of benefit that the great experimenter is abundantly justified in allowing its fame to be spread through all lands, in order that it may be tried on as large a number of unfortunate victims of dog-bite as possible. It is also clear that there is not the slightest warrant for those who would pronounce an adverse judgment on Pasteur's treatment and compare him to the quacks who deal in "faith-healing" and such like methods.

What is above all things desirable at the present moment is, that thorough and extended researches should be made by independent scientific experts in this country on the lines travelled over by M. Pasteur. This, alas! is impossible. Our laws place such impediments in the way of experiments upon animals, that even a rich man, were he capable, could not obtain the licenses necessary for the inquiry ; and secondly, the men who are most likely to be capable of inquiring into the matter are not in a position to give up the whole of their time to it, and to pay competent assistants. No one in this country is given a salary by the State, and provided with laboratory and assistants, for the purpose of making such new knowledge as that by which Pasteur has brought the highest honour to France and inestimable blessing to mankind at large. On the other hand, it is in consequence and as the direct result of such a position that Pasteur has been able to develop his genius.

Pasteur himself has not explained what theory he has formed as to the actual nature of the virus of rabies, and as to the way in which his inoculations act, so as to protect an animal from the effects of the virus, even *after* the virus has been introduced into the system. Possibly he has no precise theory on the subject, but has arrived at his results by an unreasoned exploring method of experimentation. Such a method is not permissible to the ordinary man ; but in the hands of a great thinker and experimentalist it sometimes leads to great results. Charles Darwin once spoke to the present writer of experiments, not dictated by any precise anticipation of a special result, but merely undertaken "to see in a general way what will happen" —as " fool's experiments," and added that he was very fond of such " fool's experiments," and often made them. When the individual who occupies the place of the "fool" is a man saturated with minute knowledge of the subject on which the experiment is to be tried, it is likely enough that, unconsciously, he frames hypotheses here and there without taking note of what is going on in his own mind, and so is unable to state clearly how he came to make trial of this or that experimental condition.

Whether Pasteur has worked in this way, trusting to the instinct due to his vast experience, or whether he has reasoned step by step, we do not know. It is nevertheless possible for the bystander to consider the various theories which may be regarded as tending to

explain the results obtained by Pasteur in the cure of
hydrophobia.

The general fact that the ill effects of some diseases
due to specific virus or poisons can be averted by inoculating a patient with the virus in a *modified condition*—as, for instance, when vaccination is used as a
preventive of small-pox in man—may be explained
more or less satisfactorily by three different suppositions. The first supposition is that the virus is a living
matter which grows and feeds when introduced into
the body of the inoculated animal, and that *it exhausts
the soil*—that is to say, uses up something in the blood
necessary for the growth of the virus; accordingly,
when the soil has been exhausted by a modified and
mild variety of the virus, there is no opportunity for
the more deadly virus, when it gains access, to feed and
multiply. A second supposition is that the virus does
not exhaust the soil, but as it grows in the animal
body produces substances which are poisonous to itself,
and these substances, remaining in the body after they
have been formed there by a modified virus, act poison-
ously upon the more deadly virus when that gains
access, and either stop its development altogether or
greatly hinder it. An analogy in favour of this supposition is seen in the yeast plant, which produces alcohol
in saccharine solutions until a limited percentage of
alcohol is present, then the alcohol acts as a poison to
the yeast plant, and neither it nor any other yeast
plant of the kind can grow further in that solution.
A third supposition is that, whether the virus be a
living thing or not, the protective result obtained by

introducing the modified virus into the body of an
animal is due to the education of the living proto-
plasmic cells of which the animal consists. If you
plunge a mussel from the sea into fresh water, making
sure that its shell is kept a little open, the animal will
be killed by the fresh water. But if you treat the
mussel first with "modified" fresh water—that is,
with brackish water—and then after a bit introduce
it to fresh water, the fresh water will have no injurious
effect, and the mussel may be made to permanently
tolerate fresh water. So too by commencing with
small doses, gradually increased, the human body may
be made to tolerate an amount of arsenic and of other
poisons which are deadly to the uneducated.

Any one of these three suppositions would at first
sight seem to offer a possible explanation of the protective inoculation against rabies and hydrophobia.
It is not known that the virus of rabies is a separate
parasitic organism ; at the same time it is possible that
it is. If it is not, the last of the three above-named
hypotheses would seem to meet the case, and, whether
the virus is a living thing or not, has an appearance
of plausibility.

But how are we to suppose that the inoculation
of modified rabbit's virus acts upon a man so as
to cut short the career of a dog's virus which has
already been implanted in the man's system by a
bite ?

To form any plausible conception on this matter
we ought to have some idea as to the real significance
of "the incubation period," and this we are not yet

able to form satisfactorily. Most diseases which are propagated by a virus—as, for instance, small-pox, scarlet fever, typhoid, syphilis—have a fixed and definite "incubation period." What is going on in the victimised animal or man during that incubation period? On the supposition that the virus is a living thing, we may imagine that the virus is slowly multiplying during this period, until it is sufficiently abundant to cause poisonous effects in the animal attacked. It is difficult to suggest an explanation of the incubation period if we do not assume that the virus is a living thing which can grow.

The poisonous effects are, at any rate, deferred during this incubation period. If you could introduce a modified and mild form of the same virus with a shorter incubation period into the animal which has been infected with a stronger virus with a long incubation period, you might get the protoplasm of the infected animal accustomed first to mild and then gradually to stronger doses of the poison before the critical period of the long and strong virus arrived ; and so, when the assumed hour of deadly maturity of the latter was reached, the animal tissues would exhibit complete indifference, having in the meantime learnt to tolerate without the slightest tremor of disorganisation the poison (or it may be the vibration !) which, previous to their education, would have been rapidly fatal. Almost equally well we may figure to ourselves the state of preparation brought about if we choose to employ the terms of the first or of the second supposition above given. The point of importance to ascertain, if such a conception is to be applied to Pasteur's treatment of hydrophobia, is whether the dog's and wolf's virus is longer in incubation and stronger in poisonous quality than that of the rabbit's cords as modified by hanging up in dry air. A general principle appears to be—according to M. Pasteur—that, in regard to rabies, the *longer* the incubation period the *less* the virulence of the virus, and the *shorter* the incubation period the *greater* the virulence. The virus in the cord of the rabbits used by M. Pasteur for preventive inoculation is stated by him to be, when fresh, much more intense than that taken from a mad dog; it produces rabies in a dog, when injected into its veins, in eight or ten days. By hanging in dry air for a fortnight this cord loses its virulence. But it has not yet been stated by Pasteur what are the indications that this virulence is lost, and whether the loss of "virulence" is in this case measured by an increase of incubation period. We have no information from Pasteur on this point. It would certainly seem that the virus of the dried rabbits' cords ought not to lose its short incubation period if it is to get beforehand with the dog-bite virus, which has a period of five or six weeks.[1] And presumably, therefore, there must be two distinct qualities in which the

---

[1] The incubation period of five weeks ordinarily observed in the case of men bitten by rabid dogs may be due to the *smallness* of the dose, since Pasteur has shown that small doses of rabid virus give longer incubation periods than large doses. How far a dose of weakened virus can be made to attain the rapid action of strong virus, by increasing the quantity of the weaker virus injected, has not been stated by Pasteur.

virus can vary : one, its incubation period, and the
other its intensity of action, apart from time, but
in reference to its actual capability or incapability of causing disease in this or that species of
animal.

It is useless to speculate further on the subject at
present.   The secret is for the moment locked in
Pasteur's brain.   Had we in this country a State
Laboratory or any public institution whatsoever in
which research of the kind was provided for, the
fundamental statements of Pasteur as to his results
with dogs would ere this have been strictly tested
with absolute independence and impartiality by
English physiologists retained by the State to carry
on continuously such inquiries.   Similarly, we should
have independent knowledge on the points above
raised as to the modification of the virus in rabbits,
and the public anxiety on the whole matter would be
in a fair way towards being allayed.   At the same
time, in all probability similar treatment in regard to
other diseases would ere this have been devised by
" practical " English experimenters.   As it is, owing
to our repressive laws and the State neglect of scientific
research, we have to remain entirely at the mercy of
the distinguished men who are nurtured and equipped
by the State agencies of our Continental neighbours.
All that we are in a position to say with regard to
Pasteur's treatment of hydrophobia is, that unless the
accounts which have been published in his name and
by his assistants are not merely erroneous, but wilful
frauds of incredible wickedness, that treatment is

likely to prove a success so extraordinary and so beneficent as to place its author in the first rank of men of genius of all ages. That is the position, and there is no reason why the former alternative should even for a moment be entertained.

APPENDIX

THE period of two years and a half which has elapsed since the preceding article was written has extended our knowledge of the results of Pasteur's system of inoculation, and has also given more definite shape to the theories as to the *modus operandi* of the protective virus.   Up to the commencement of last summer 7000 persons had been inoculated at the Institut Pasteur in Paris.   The average mortality of these individuals was 1 per cent, or, in other words, there were 70 deaths.   If the mortality had been 15 per cent—the percentage of deaths of persons bitten by rabid dogs, as nearly as can be estimated, before Pasteur's treatment was known—then there would have been as nearly as possible 1000 deaths ; so that we are justified in saying that Pasteur has already saved in this one institute alone 900 lives.   Even if we make a reduction in the number of cases treated by considering those only in which the dog which bit was certified to be rabid by a veterinary surgeon, or experimentally proved to be so by inoculations made after its death with its spinal cord, we have still 4500 cases, with a mortality of 70 instead of a mortality of 675.   In various places in Russia and Italy, and in Turkey and Havana, Pasteur's system has been introduced with the same satisfactory results, viz. a reduction of the mortality of persons bitten by rabid dogs from 15 per cent to 1 or 2 per cent.

The number of inoculations with the preventive virus and the strength of the virus used has been gradually modified during the past two years, in such a way as to render the later results better than those at first obtained. In fact, the latest returns show a reduction of the mortality to less than 1 per cent, and in some series to zero — as, for instance, at Warsaw, where Dr. Bujarid inoculated in 1887-88 no less than 370 persons, all bitten by animals proved to be rabid ; of these persons not one died. A similar result was obtained at Palermo by Dr. A. Celli, who in 1887-88 inoculated 609 persons.

The following fact is perhaps the most convincing in regard to the efficacy of Pasteur's treatment. In the year 1887 the official medical report shows that 350 people were bitten in Paris by animals suffering from rabies. 306 of these persons were inoculated at the Institut Pasteur ; of these 3 died. 44 of the 350 declined to be inoculated, and of these 7 died. We have here the two groups—the inoculated and the non-inoculated—drawn indifferently from the same population, living under the same average conditions, and subjected to treatment at the same time. The results are precisely confirmatory of the conclusions arrived at from the study of the larger series of figures derived from less rigidly established data. The percentage of the deaths of those who refused to be inoculated is 16 ; the percentage of the deaths of those treated by Pasteur is 1.

It is a matter of great satisfaction to me to be able to record here the recognition of the convincing character of these results by a large and influential body of representative Englishmen. The Lord Mayor, Sir James Whitehead, kindly undertook, at my urgent request, to visit the Pasteur Institute when he was in Paris in May 1889. What he saw made so favourable an impression that he readily undertook, on his return, to do something to show the gratitude of the people of this country towards M. Pasteur, by whom 250 British subjects had already been treated, and he consented to call a meeting at the Mansion House if I would endeavour to secure the co-operation of the leading members of the medical profession, of the President and Council of the Royal Society, and other representative men. The meeting, which was held on 1st July, has been declared to be the most influential which ever assembled in the Mansion House. Among those who were present were the Duke of Northumberland, the Duke of Westminster, Sir John Lubbock, M.P., Sir James Paget, Sir William Bowman, Sir Spencer Wells, Sir Henry Roscoe, Sir Joseph Lister, Sir Trevor Lawrence, M.P., Colonel Sir Edmund Henderson, General Strachey, the Dean of Manchester, Prebendary Harry Jones, the Vicar of St. Pancras, the President and Council of the Royal Society, and deputations from various medical societies and from dog-fanciers' associations. The Prince of Wales, Professor Huxley, Professor Tyndall, and the

French Ambassador wrote letters warmly supporting
the objects of the meeting. The following are the
resolutions which were passed :—

(1) That this meeting records its conviction that
the efficacy of the anti-rabic treatment discovered by
M. Pasteur is fully demonstrated.

(2) That this meeting desires to express the in-
debtedness of the people of Great Britain and Ireland
to M. Pasteur and the staff of the Institut Pasteur for
the generous aid afforded by them to over 200 of our
fellow-countrymen suffering from the bite of rabid
dogs.

(3) That this meeting requests the Lord Mayor
to start a fund for the double purpose of making a
suitable donation to the Institut Pasteur, and of providing for the expenses of British subjects bitten by
rabid animals, who are unable to pay the cost of a
journey to Paris.

(4) That this meeting, whilst recognising the
value of M. Pasteur's treatment, and taking steps to
provide for the treatment of persons who may hereafter be bitten by rabid animals in this country, is of
opinion that rabies might be stamped out in these
islands, and invites the Government to introduce without delay a bill for the simultaneous muzzling of all
dogs throughout the British islands, and for the establishment of quarantine for a reasonable period of all
dogs imported.

The fund which was there and then started has
enabled the Lord Mayor to assist many poor British
subjects to visit Paris for the purpose of undergoing

M. Pasteur's treatment, and has further enabled the committee to make a donation of 30,000 francs to the Pasteur Institute.

With reference to the theory of preventive inocu-lation for hydrophobia, it must be remembered that the whole subject is one of the most recent develop-ments of biological doctrine. The fact was long ago established that inoculation of a disease may be suc-cessfully practised for the purpose of producing a *local* development of the disease, which acts so as to protect the organism against a more general attack supervening by the ordinary channels of infection. Such inoculation was practised as a protection against small-pox before the discovery of Jenner, and has also been used in relation to some other infective diseases. It may be spoken of as "Montaguism," in reference to Lady Mon-tagu, who introduced the practice into this country. The use of the disease in a modified form for the pur-pose of inoculation (or of a distinct but allied disease for this purpose) is what is known as Jennerism, and is seen in the use of the cow-pox as a protective against small-pox. The further step of taking the infective material, which is nothing more nor less than living Bacteria, or microbes, of a special kind, and cultivating it artificially in solutions with increased oxygenation · or temperature, or whatever it may be, so as to diminish its poisonous quality, and then using this cultivated material for preventive inoculation, is Pasteurism. It has been most extensively applied in the case of the splenic fever, or anthrax, of cattle and sheep, but also in many other diseases experimentally.

It is now demonstrated that the protection afforded, whether by Montaguism, Jennerism, or Pasteurism, is due to the production of a chemical soluble substance by the microbes of the disease, which may be termed generically "vaccin." The vaccinated animal or man becomes impregnated with this vaccin, and is no longer liable to suffer when the microbes of the related disease enter the blood or tissues by chance infection. The protection cannot, however, be due to the accumulation and retention of the vaccin in the animal's body, for it has been shown by Bouchard that vaccin substances are excreted in the urine, which thus acquires vaccinating qualities. It must therefore be due to an action on the protoplasm of the living cells of the organism, by which the molecular condition of the protoplasm is slightly but permanently, or at any rate for a considerable time, changed. The condition of the chemical activity of the protoplasm—summed up as "nutrition"—is affected, and it is no longer liable to suffer from the poisonous activity of the microbes related to this particular vaccin.

An extremely important question is whether the vaccin produced by a microbe is the *same* substance as the poison by which it acts so as to produce deadly effects. The poisonous product may be called the "toxin." It is undoubtedly, like the vaccin, a soluble, diffusible, chemical body produced or secreted by the microbe. It is quite possible that in some cases the toxin and the vaccin may be one and the same substance, in which case the protective action would be simple mithridatism, as explained in the note at p. 113. But in other cases the toxin and the vaccin seem almost certainly to be distinct. From what we know of the complexity of the chemical changes and the variety of substances produced by Bacteria and other schizophyta in fermentations set up by them elsewhere than in a living animal's body, it would not be surprising if eventually we should be able to discover and isolate in the case of each kind of pathogenic microbe several toxins and several vaccins. The great object of this line of research in the future will be to obtain by artificial culture of the microbe, and subsequent treatment of the products of its life, the vaccin of each infective disease in its chemically pure condition, free from admixture with any of the toxin, and free from the microbe itself. We have to subdue and utilise these ultra-minute organisms, and to make them render service to man, as we have done with larger and even more deadly enemies. Already we employ a microbe to prepare our drink and to lighten our bread, and another to produce our vinegar, and a third to flavour our cheese. We shall hereafter be able to breed these dainty cattle and select strains suitable for our uses, making the domesticated protect us from the untrained kinds, even as the dog guards us from the wolf.

We now know from M. Pasteur himself what he had not explained in 1886, viz. that he has been guided in his work on preventive inoculations, and especially in that on hydrophobia, by the theory that the toxin and the vaccin are two distinct soluble

chemical substances produced by the microbe. In
the spinal cord of a rabic animal removed and hung
up in a glass jar to dry, the vaccin substance appears
to remain intact, whilst the toxin diminishes, and
finally is all destroyed.

As stated in the essay to which this note is an
appendix, M. Pasteur does not use in his treatment
of persons bitten by rabid animals an attenuated virus,
that is, a virus containing a weakened toxin, but, on
the contrary, a virus of increased strength. The rabic
virus as found in the dog is of medium strength and
rapidity of action; by cultivation from monkey to
monkey it can be attenuated, rendered slower and less
deadly in its action. But by cultivation from rabbit
to rabbit M. Pasteur obtains a virus of maximum
intensity, which kills in seven days, and it is this
which he makes use of in treating persons who have
been already infected by the virus of medium strength
from the tooth of a dog. The explanation of this is,
of course, related to the fact that M. Pasteur has not
in his treatment to prepare his patients to withstand
a future inoculation by dog-bite, but has to treat
persons already bitten, already harbouring in their
bodies the slowly-acting form of the virus derived
from the dog.

"It is," Dr. Bouchard observes, "necessary to
overtake this latter virus, to act more quickly than it
does, in fact to make use of a vaccinating virus of
briefer incubation, to employ in consequence an intensified virus. Inoculated beneath the skin, a rabic
virus of any kind will often give vaccinal protection;

a virus of short incubation—the virus of the rabbit—
will vaccinate, if it does not kill, before the virus from
the dog will have produced hydrophobia. If the
object had been a prophylaxis against possible future
bites, the source of the virus used would have mattered
little; but the purpose was to obtain therapeutic inoculation. It is for this reason, I think, that M.
Pasteur chose the medulla of the rabbit as his source
of vaccinal virus. He submitted this medulla (spinal
cord) to desiccation, because he had established experimentally that, whilst the fresh medulla inserted
beneath the skin kills often, the desiccated medulla
does not kill, or kills very rarely. Further, he made
out experimentally that the more of the medulla he
inserted, the less chance there was of killing and the
greater chance of conferring immunity. In consequence, M. Pasteur found it necessary to multiply his
inoculations, to make several every day and for several
days. Guided, no doubt, by prudential reasons of a
purely theoretical character, he chose for the first
inoculations the oldest medullas, which are no longer
virulent, or are very nearly ceasing to be so, and he
then passed on to medullas of a little more dangerous
character, thinking thus to establish step by step
partial immunities, so as to render each subsequent
inoculation with a less desiccated medulla quite in-
offensive. I imagine that the method of anti-rabic
vaccination was thus built up little by little under
the influence of theoretical views and experimental
conclusions. The practice having been demonstrated
good by experience, its formula has remained, and

has not varied, even although the theory has been modified profoundly.

" At the present moment M. Pasteur believes that his desiccated medullas contain a virus, not attenuated or altered in *character*, but diminished in *quantity*, and that it is less abundant in proportion as the medullas have been long kept. He considers that this virus will require a longer time to produce hydrophobia than would a fresh medulla, because the duration of incubation is *ceteris paribus* proportional to the quantity of the virus. He supposes that during this long incubation there is an advantage in that the vaccinal effect—immunity—has time to produce itself, but he explicitly states that this immunity is not due to the virus. It is, he says, produced by a vaccinating substance which exists in the medulla side by side with the virus, and that desiccation, whilst destroying the virus (toxin), leaves the vaccinating matter (vaccin) intact.

" Consequently the virus has nothing to do with the result ; in the anti-rabic vaccination there is only one thing of any service, and that is the vaccinal matter. That being so, why choose an intensified virus, a virus of short incubation ? Why make successive inoculations with medullas of increasing virulence ? Why not limit oneself to a medulla of fourteen days' desiccation—a medulla which has ceased to be virulent, and is still vaccinal, which contains what is useful and nothing which is risky or dangerous ? "

To these questions M. Bouchard replies that, probably because the present system has been found

efficacious, there is a natural reluctance to change
any of its empirical details, but that further experiment with animals may lead to modification in the
direction indicated by his queries.   On the other
hand, it is admitted by all who have occupied themselves with the subject that there is still a vast deal
to be ascertained experimentally in regard to the
whole question of vaccins and immunity.   A curious
commentary on M. Bouchard's anticipations is found
in the fact that the Russian physicians have modified
Pasteur's system in the opposite direction.   They
carry on the series of inoculations till medullas of
only two days, and in some cases fresh medullas not
twenty-four hours old, are reached.   The results of
this intensive treatment are said to be extremely
satisfactory, giving less than 1 per cent of deaths.
The *Thérapeutique des Maladies Infectieuses, Anti-
sepsie* of Professor Bouchard, published in Paris in
1889, contains the most luminous exposition of the
whole subject of vaccination and immunity in reference to Bacteria with which I am acquainted.   Those
who may have been stimulated by the foregoing
remarks to a desire for further information on these
topics cannot do better than peruse M. Bouchard's
admirable work.

IV

EXAMINATIONS

FROM THE *UNIVERSAL REVIEW*, NOVEMBER 1888

EXAMINATIONS

I DO not feel much diffidence in complying with the request of the Editor of this Review, that I should write a brief statement of my opinion on the recent attack upon examinations, because I believe that there are very few persons who have had so wide and continuous an experience in examinations as I have had. In early days I was subjected to innumerable examinations for scholarships, degrees, and fellowships, both at Oxford and at Cambridge, and during the past twenty years I have been almost constantly employed in examining for scholarships, or degrees, or fellowships, or other appointments on behalf of the University of Oxford or of that of Cambridge, or for my own or another college, or for the University of London, or for the Civil Service Commissioners, or the Public Schools Examination Board.

A mere experience of a vast number of examinations would not, I think, in itself enable any one to form an opinion as to the good or evil results flowing from the present examination system were that experience not accompanied, as it has been in my own case, by constant occupation as a teacher: I have been thus occupied both at Exeter College, Oxford, and at University College, London. It is only by close contact as a teacher with the actual class of students who are submitted to any particular examination, that it is possible to form a really well-based judgment as to the influence which the examination system is exerting upon education, and it is only by actual experience as an examiner that a correct apprehension of the limitations and possibilities of the examination system can be gained.

Though I agree with the general statement of the recently published protest that the examination system is at the present moment exerting a most injurious influence upon the youth of this country, I did not feel able to join in signing the published paper on account of the indiscriminate character of the attack therein made.

I have often thought that in these days of congresses a conference of examiners might with advantage be brought together in order to discuss the Art of Examining, and if possible to frame some general conclusions as to the best methods of examining in various cases and for various ends, as well as to consider the utility or inutility of examinations as at present applied to a variety of purposes.

It appears to me that it may very well be conceded without discussion that it is likely that examination is useful and in every way satisfactory in regard to some of the purposes to which it is at present applied, whilst it may be productive of harm when applied in relation to other purposes.

No statement or discussion concerning examina-

tions can therefore be admitted which does not recognise this distinction; the question, for instance, of
the utility of qualifying examinations must be kept
apart from that of the utility of competitive examinations.

Moreover, it hardly admits of dispute that, whilst
some subjects of study or accomplishment can readily
be made the matter of examination, and will yield the
most conclusive evidence as to the acquirements of
an examinee when they are thus used, other subjects
are extremely difficult to examine in. Thus elementary mathematics form perhaps the ideal subject of
examination, whilst philosophy and other speculative
matter are troublesome and illusive from an examiner's
point of view. More than this, whilst it is possible
to test by examination the memory of the examinee
as to the contents of a book, and even his real knowledge of a subject, it is very much more difficult to
test by examination his possession of a variety of
mental qualities, such as administrative capacity, tact,
intellectual honesty, and other moral attributes; at
any rate no such testing, even could it be devised, has
hitherto been practised.

Hence it seems that the whole question of examination must be limited by a consideration of the
subject-matter to which examination can be applied.

A third element of very great importance in any
estimate of the influence of examinations is that of
the persons who examine and the persons who are
examined. It is maintained by many thoughtful
observers that there is no harm in examinations so

long as they are limited to the examination by a
teacher of his own pupils. It is maintained on
the other hand—or perhaps I should say was maintained some years ago—by distinguished educational
reformers, that the examination of his pupils by a
teacher is harmful and absurd, and that the only
good that can be done by examinations is done
when the examiner is a stranger alike to the examinees and their teachers. Under this head also
arises the consideration whether examinations are
at all times in all places good for all men. It may be
that examinations of one kind and another can be
advantageously applied to schoolboys, but are not
useful to the same extent or for the same purposes
when applied to university students or to professional
men. Some persons think examinations bad for
infants, whilst admitting their harmlessness when
applied to children above ten years of age; some
think that women as a rule are injured by the examination rack, which only permanently injures a
small percentage of their brothers. Accordingly, we
must note as a variable feature, which must be limited
and defined in all statements as to the influence of
examinations, the personal element sought by the
questions : Who are the examiners? Who are the
examinees? What are the relations of the one group
to the other?

It is impossible in the space at my disposal to go
over from the three points of view above stated the
whole series of examinations held under one or other
authority in this country. I will confine myself to a

statement derived from such special experience as I have had, which includes examinations of four distinct kinds, viz. :—

I. Examinations applied as part of the educational system in schools and universities.

II. Examinations qualifying for admission to a profession, in which special knowledge is a necessary equipment for the honest practice of the profession.

III. Examinations for the purpose of selecting by strict competition the one or more successful candidates from among a number of applicants for a post in the Home or Indian Civil Service.

IV. Examinations for the purpose of selecting by strict competition—the award going to the examinee who has scored most marks—the recipient of a scholarship or of a university fellowship.

I. Firstly, with regard to the use of examinations in schools and universities. I cannot admit that there is any well-founded objection to the use of examinations as a means of testing and determining the progress of a student in an educational curriculum; on the contrary, it seems to me that pass examinations at definite intervals are necessary both to enable the teacher to ascertain whether the student has gained sufficient knowledge in order to proceed to a further grade or branch of instruction, and to enable the teacher to correct his own methods if found defective. To the student also such examinations are a most valuable exercise, since by passing them he gains confidence in his own progress, and in the attempt to pass them, surveys and brings into practical form the results of study. Such examinations, however, assume a different aspect when they are conducted, as is too frequently the case, by an examiner appointed *ad hoc*, who is a stranger both to the examinee and to his teacher. A further modification which is as a rule injurious—though not, perhaps, under all circumstances—is introduced when the examination is made competitive, and the results are published to the world with the examinees' names arranged in order of merit. There is, I think, a distinction to be drawn here between the examination of schoolboys and that of university students. My strong impression is that competition by examination for pre-eminence in class among schoolboys is not injurious, but, acting upon healthy young natures, serves as an unobjectionable stimulus to exertion. It also seems to me that, whilst frequent or even annual inspection and examination by external examiners is open to the same objections in the case of schools as in that of universities (to be mentioned below), there is not the same objection to a "leaving examination" or "university matriculation examination" being conducted by examiners who have not been the teachers of the candidates. Such an examination as that last named may be regarded as a means of criticising and testing the performance not merely of the schoolboys but of the schoolmasters. And it is right that they should be so tested and directed by those charged with the later steps of education, namely, the authorities of the university. The university must necessarily be regarded as the

highest authority in the educational system, and it is
desirable that it should give direction to school education, and maintain by its criticism a certain standard
of efficiency in the work of masters and scholars alike.
On the other hand, there is no authority beyond that
of the university, and accordingly there is not the
same reason for subjecting university students to
examination by persons who have not been their
teachers as there is in the case of schools.

The plan of setting "impartial" examiners who
have had nothing to do with the teaching of
the examinees to test them by written papers and
vivâ voce questioning is the primary evil from which
all the abuse of examination in the English universities
springs. The system has grown to such dimensions
and acquired such collateral developments in Oxford
and Cambridge that in my opinion the condemnation
expressed by Professor Freeman and Mr. Frederic
Harrison is fully justified. The system administered
by the examining board in Burlington Gardens, known
as the University of London, is even more injurious in
its results. I venture to express this opinion as
having acted for some years in the capacity of examiner for the University of London and at intervals
for the Universities of Oxford and Cambridge. It
is desirable to state a little more fully what the
system objected to is, and the nature of the evil
which it produces. The special examiner—if I may
use that term to signify an examiner who is not
the teacher of the examinee—is employed both to
carry on pass examinations, which in themselves are

free from objection, and indeed necessary parts of a university curriculum, and also to conduct competitive examinations for "honours" and university prizes. He examines pupils taught in a number of different colleges, who compete with one another like the race-horses from rival stables. The teachers act the part of trainers, and university teaching assumes the form of a keen and restless competition for success in the examination. The result in regard to the teaching is most disastrous. It has been deliberately maintained by the advocates of the system (to carry out which the present University of London exists) that its chief advantage is that by employing first-rate examiners you can dispense with any provision for first-rate teachers, or indeed any teaching at all, since the examining body lays down in detail the subjects of the university curriculum, and by a free competition those teachers will come to the front who succeed in passing most candidates or gaining most honours. The direct result of this plan is that in London there are a large number of teachers who take very low fees, and who teach their subjects solely with the view of securing a pass for their pupils with the minimum expenditure of time. Competition leads to the underselling of capable men by inferiors, and a consequent diffusion of the students among a number of small private or semi-public institutions where, as the results of the examination-room show only too clearly, they are not well taught. At Oxford and Cambridge the result of the system of special examiners is modified by the "peculiar institution" of those universities—

the colleges. The competition is not amongst individuals or small impoverished schools, but amongst
wealthy colleges. Accordingly the teaching never
sinks to so low an ebb as in London ; college endowments and the college monopoly of fees enable the
Oxford and Cambridge colleges to employ men of
ability and culture as "trainers" for the examination
stakes, and each college competes with its neighbours.
But a curious result of the combined system of college
monopoly and special examiners is that the university professors are, in regard to most subjects taught
in colleges, left without occupation, having, as a
rule, very few pupils and no voice in the examinations.

The result of these arrangements, whether in
Oxford and Cambridge or in London, is that the
examinations acquire an undesirable importance. A
man refers throughout his life to the fact that he
obtained a "first-class" as a sort of perpetual testi-
monial, and as the complement of this attitude the
examination is insensibly, and without deliberate
intention, made more difficult every year. The men
who have been themselves examined in this way, and
who have taught in connection with this system, in
due course become examiners or assist in preparing
schedules of the examination, and each gives a turn to
the screw until the condition is arrived at when a
candidate for honours has no time to read or to think,
but must rush, note-book in hand, from one skilful
lecturer or coach to another, who undertake to put
him through "all that is necessary for the examina-

tion," in so many hours a week, in twenty different branches of study.

Perhaps the most injurious result of the system is the degradation of the teacher. There is no nobler vocation among men than that of the teacher; there is no more enviable position in the world than that of the university professor as realised in Germany— honoured by the State, chosen from among his com- peers to pursue his studies as a public servant and to associate with himself in the making of new knowledge the best youth of the Fatherland. For him there is no intrusive board of examiners—drawing away from him the attention and respect of his pupils, or urging him to put aside his own thought and experience, and to teach the conventional and commonplace. The system invented by Fichte maintains that perfect relation of teacher and pupil which he sketched in his description of the University to be founded at Berlin in the beginning of this century.[1] The essence of that relation is the absence of examiners—the professor himself is examiner and teacher in one. Another important feature is that the professor is neither dependent on fees, nor entirely independent of them. He has a stipend which ensures him a certain freedom of action, and his pupils pay him a small fee which

---

[1] Fichte says in so many words that a university is not a place where instruction is given but an institution for the training of experts in the art of making knowledge, and that this end is attained by the association of the pupil with his professor in the inquiries which the latter initiates and pursues. Such a university is what very many Englishmen who have studied in Germany desire to see in London, in Oxford, and in Cambridge.

ensures his attention to their requirements as well as their respect for what is not merely given for nothing. Compare this with the position of the professor who teaches under the auspices of the University of London, or even with that of the college lecturer of Oxford or Cambridge. The university, in order to carry out its system of examinations, is obliged to produce a detailed schedule of every subject in each examination, and it is only in accordance with this schedule that the professor or lecturer can open his mouth. Such schedules necessarily become antiquated; they are often futile and objectionable even when freshly prepared by the "boards" of examination; but the most gifted teacher is tied down to these schedules, drawn up often by those who have passed from a former period of "activity" to a later and mischievous period of mere "authority." The pupil necessarily under this system looks upon his teacher as an inferior person, who like himself is subject to the dictation of examiners; he resents the introduction of any matter in the course of the teaching offered to him which is not in the schedule, or will not pay in the examination; he pays his fee not to study a subject, but to be "put through the examination," and, under the system of open competition among teachers which exists in London, he tends to seek that teacher who most frankly accepts the odious position of the examination hack.

In my own opinion, as having had experience of it at University College, the attitude thus forced on pupil and teacher by the examination system actually

at work is most pernicious to the intellectual development of the student, and to the teacher almost un-
bearable. Fortunately for me my experience has been
largely mitigated by the fact that I have independently held the post of examiner at Burlington House
during the greater part of the time in which I have
been teaching at University College. And this
brings me to a curious and important consideration.
The advocates of the examinations of the University
of London claim, and from the first have claimed,
that their great recommendation—as contrasted with
the examinations of a German or a Scotch university,
in which there is no schedule and the professor is
examiner—is that in the London system the schedules
are laid down by an independent board of learned and
impartial authorities, who are not teachers, and that
the examinations are conducted by highly-paid
examiners who have no interest in or knowledge of
any of the candidates; therefore (it is said) the results
deserve the very greatest confidence—they are absolutely impartial. Now as a matter of fact the world
is so constructed that this pretence of impartiality is
sheer nonsense. A superior class of beings who can
draw up schedules and examine without teaching does
not exist. The examiners of the University of London
are, as a matter of fact, teachers; and often, as in my
own case, a fourth of the candidates in an examination
have been taught by the examiner, whilst the remaining three-fourths suffer the disadvantage of not having
been so taught.

A final objection to the " special examiner " system,

which is perhaps of more practical value than any
other objection, is this, viz. that one of the real
advantages of examination in connection with education is lost when a stranger examines instead of the
teacher. That advantage is the observation on the
part of the teacher of the failure or success of his
methods of teaching. The importance of this experience to the teacher cannot be exaggerated. Where
external examiners act they necessarily give most
meagre reports as to the performance of the examinees;
" reject," "pass," "honours," are the actual words which
convey all the information vouchsafed to the teacher.
At the University of London no proper reports on
the performances of unsuccessful candidates are ever
given, and no use is made of the examination for the
purpose of correcting the teaching in the different
places of study which supply candidates. The reason
for not giving such reports is adequate and is condem-
natory of the whole system. It is this : There are so
many candidates that it would be impossible to furnish
a full report of each candidate's work to his teacher.[1]

The opinion then which I hold in regard to examinations in the universities is that they are excellent
and useful only on the condition that they are con-

[1] The defenders of the examination system of the University of
London are in the habit of quoting the extremely large numbers of
its candidates as evidence of the success of the university. On the
other hand, it seems that really these large numbers are one of the
strongest objections to the system pursued—since it is *impossible* to
maintain uniformity of judgment with more than one hundred
examinees, or to give proper time and attention to the preparation of
reports on the work of each examinee which may serve as guides to
the teacher in each case for the future.

ducted by the teacher, and that they are not competitive and have no honour or distinction assigned to them, but are used simply as pass-tests for admission to the degree. The teachers in every university ought, I venture to think, to be professor-examiners, and the student should be admitted to his degree whenever he has obtained the certificate of having passed the class examination of three of such professors. It is not impossible that at some future time the college-lecturers of Oxford and Cambridge will be grouped with the university professors, each representing a special subject or branch of a subject, and such an organisation as that required to make the teaching and examining in these universities professorial may be brought about. That time, however, is undoubtedly remote. In London there is perhaps a better chance of the establishment of a professorial university : it is a chance depending on the report of a Royal Commission now sitting. There is no reason why there should not be more than one University in London, it is big enough to occupy several.

II. With reference to examinations qualifying for admission to a profession, there can, it seems to me, be no question as to their necessity and their harmlessness, if conducted with certain restrictions. Such examinations ought to be administered by some high authority representing either the profession or the State. The examination should be a single qualifying examination in that knowledge necessary for the practice of the profession. There should be no attempt to control the steps of the education of

the candidates; the examination should be in the
minimum of necessary subjects, and the standard of
passing should be the lowest compatible with the
safe guarding of the public. No competition for
honours or distinction of any kind should be associated
with such examination. The education of candidates
for the professions—the church, the law, medicine, the
army—is and should be undertaken by bodies independent of the profession. The university degree, or
the Fellowship of a College of Surgeons or Physicians
should be altogether independent of and an addition
to the possession of the State or professional diploma.

III. The question in regard to examinations,
which undoubtedly has the greatest interest for the
general public, is that as to the desirability of continuing to employ them for the purpose of selecting
appointees in the Home and Indian Civil Service. I
must admit that, strongly opposed as I am to competitive examination in the universities, I do not see
how to avoid it in this matter. I believe that a great
boon was conferred on the people of this country when
Government appointments were thrown open to them
in this way, and I should rather see further progress
in this direction, so as to include more valuable
offices, than a retrograde movement. The whole
question here seems to me to limit itself to this :
"How to improve the Civil Service Commissioners'
examinations so that ability and qualities of the kind
desired for any given post may be selected as the
result of those examinations." I have no doubt that
a great deal might be done in the way of improving

these examinations. In some cases it is easy to adapt the examination to the work required of the future Civil servant, and then if the examination be long and thorough I do not see what better system of selection can be invented. Thus, for instance, in selecting from among the gentlemen who apply to be appointed to vacancies on the staff of the British Museum, it is easy to apply an examination which is directed to the subject with which the future curator will be occupied. This method has been adopted not so long since, and its results seem to be satisfactory. It would be possible to adopt a similar style of examination in regard to other offices, and even the candidates for Treasury clerkships might be examined in such a way as to test, not their preparation, but their general intelligence and mental agility. Probably the effort which has been made to throw the Indian and other Civil Service appointments into the hands of the upper classes by favouring in the examination the subjects taught at the public schools, is chiefly to blame for the apparent absurdity of the examination system as a test of a man's fitness to act as an Indian magistrate or an inspector of primary schools. The examinations might be adapted—no doubt at the expense of some trouble and ingenuity— to the special requirements of each kind of appointment.

IV. Space does not remain for me to discuss, as I should like to do, the subject of competitive examination for scholarships and for fellowships. Here the competition falls within the school and university curriculum, and is accordingly objectionable. The competition for university entrance scholarships amongst the large schools, and the glorification of the school with the longest list of successes of this kind, are bad. School education becomes, in consequence, more and more confined to those subjects which lend themselves readily to examination competition. Boys are specialised in their studies too soon, and the clever boys who have a chance of scholarships are over-taught, whilst the stupid boys, who need more teaching, are neglected. Besides this, the awarding of these scholarships to rich and poor alike by open competition simply increases the average expenditure of the university undergraduate by £80 a year. Those whose parents can pay for the best coaching, that is to say, are richest, carry off these prizes from the poorer competitors, and what should be (and in some cases is) the means of supporting a youth at college, becomes simply so much extra pocket-money. I would award these scholarships, after careful inquiry, to those who were really unable to exist at the university without them. A special organisation would be required for their administration, designed so as to avoid jobbery and nepotism. The college which awards the scholarship might require a legally certified statement of poverty from each candidate, and a limited competition by examination, not in special subjects, but in a wider range than is at present usual, might then be applied. The stigma of poverty thus attaching to the "scholar" would do no harm, but good. There is too little poverty and too much luxury among Oxford and Cambridge undergraduates at present.

With regard to fellowships—the plan which has been adopted within the last ten years both by Oxford and Cambridge colleges with reference to the award of fellowships for distinction in natural science might well, I think, be adopted in regard to fellowships in all subjects of study. Candidates for the fellowships are requested to send, in support of their application, copies of any original works which they may have published. There is no general examination, but a few questions may be set to each candidate with special reference to the matter of his original work. The work done by a candidate at his leisure, and not the gymnastic performance in an examination room, thus decides the competition. Fellowships have been awarded in this way at Trinity, St. John's, Christ's, and Clare Colleges at Cambridge, and at Lincoln, University, and New Colleges at Oxford.

V

THE SCIENTIFIC RESULTS OF THE INTERNATIONAL FISHERIES EXHIBITION, LONDON, 1883

A Lecture delivered at the Exhibition in June 1883

THE SCIENTIFIC RESULTS OF THE INTERNATIONAL FISHERIES EXHIBITION, LONDON, 1883

THE text which has been selected for the Paper which I have the honour to submit on the present occasion has caused me no little perplexity on account of its ambiguity.

It has been pointed out to me that it is unwise to prophesy unless you know, and that no one at present can know what may be the results, scientific or otherwise, of the great Exhibition, which has still some months of its career to run.

Again, it is apparent that the word "scientific" has a very wide scope, including statistical, mechanical, hydrographical, biological and sociological results, all of which are in some way or other to be observed and studied in the great International Fisheries Exhibition.

The comprehensive vagueness of the title of my discourse has, on the other hand, the advantage that it permits me to choose from a very wide range of subjects, and I have accordingly to submit to you the following as a more exact definition of the matter to which I desire to call your attention. I propose, not to speak so much of scientific results which may flow from the Exhibition, as of scientific results which are illustrated in the Exhibition, and in particular of those results of the science of Zoology which are of importance to the Fish Industry, and are more or less completely set forth for our instruction and edification in the collections which have been brought together in the London International Fisheries Exhibition.

It would have been a congenial task to me to describe here some of the rare specimens of great interest to the zoologist, which have been sent by foreign countries to this exhibition. Such specimens as Nordenskjold's Rhytina and the magnificent skeletons of Xiphioid Whales shown in the Swedish Court are of surpassing interest and importance from the zoological point of view. At the same time it must be admitted that they do not have any special importance in relation to Fisheries, and accordingly I must leave unnoticed such rarities and delights of the zoologist, in order to address myself more especially to the question of the relationship of the science of zoology to the fish industry.

The value of zoological science in relation to fisheries is not, I think, so fully appreciated in this country as is desirable in the interests of the public, and of those who make profit by enterprise in fisheries.

There is a very general tendency among men whose occupations are of a commercial character to undervalue the work of scientific inquiry, not only in regard to such matters as fisheries and fish-culture, but also in relation to manufacturing industries, agriculture, mining, and even in relation to medicine. To a large extent this arises from a misconception as to the real nature and character of what is called "science." Science is the knowledge of causes; its method and purpose when strictly pursued lead to the accumulation and arrangement of thorough and accurate knowledge of any given subject to which it may be applied, with a certainty and an abundance which no other method and no other purpose can give. Undoubtedly the latest scientific knowledge of a subject is very usually not *immediately* useful to those who are engaged in applying commercial enterprise to the same subject. It is however to be noted, over and over again, that the scientific discovery of one generation becomes the necessary foundation of some valuable commercial enterprise in the next: what was at one time a curiosity and of little interest, save to men of science, becomes after fifty years the pivot of some great industrial manufacture.

Accordingly commercial men, and those who place the material well-being of this country beyond all things as an object to be continually striven for, should have patience in the presence of what seem to be the useless accumulations of knowledge; they should have faith in the ultimate utility of science, for already throughout the length and breadth of the land this cause-reaching knowledge, which we call "science," has proved its enormous power of aiding commerce, and has amply established its claim not to mere toleration but to eager and generous support

from those who are reaping golden harvests through
the science of a past generation.

When we remember that science is really no more
nor less than such accurate and full knowledge of this
or that class of natural things as enables us actually
to understand " the causes of things," then it becomes
obvious that the distinction which is sometimes drawn
between the "scientific" man and the "practical"
man is founded upon some kind of error.   If there is
the antithesis which fashion causes many persons to
assert as existing, let us see what becomes of it when
we say, as we are justified in saying, that the scientific
man is the man who knows thoroughly and accurately.
The contrast insisted on between the scientific and
practical man becomes, then, simply the contrast
between the man who knows and the man who does
not know, but acts in ignorance.

As a matter of fact there is no such antithesis.
Your man of science is, or should be from the nature
of his pursuits, more thoroughly practical than any
one who affects to despise scientific knowledge, for he
is accustomed to ensure success in his experiments
and investigations by taking every means in his
power to that end; above all, and chiefly, by guiding
himself by reasonings based on the most accurate
and extensive knowledge.   So too, indeed, every
so-called practical man who is not a mere adventurer
—a happy-go-lucky tempter of Fortune—makes
use of accurate knowledge to aid him in his commercial ventures and speculations; so far as he can
get it, he makes use of science, though he often calls

it by some other name as soon as it becomes useful knowledge.

The fact is, that a large part of the indifference to science in this country, and the notion that science is dreamy, vague, untrustworthy and useless to practical men, has arisen from the fact that these worthy practical men have very often allowed themselves to be imposed upon by mere quacks and pretenders, who assume the language and authority of science without any credentials whatever, and lead the practical men astray. Such quackery in science has been by no means unusual in this country, owing to the almost complete destitution of the wealthy classes in respect of scientific education. Practical men have, as a rule, not even a smattering of scientific training, and cannot distinguish true from false science, cannot tell which is the quack and which the man of real knowledge. Equally unfortunate in this respect, in former times, have been the members of the executive and delibera-tive branches of our successive governments, so that—in days which we may hope are past—ignorant pre-tenders to scientific knowledge have been, in good faith, placed in responsible positions, and have helped to justify the notion that modern science is a wind-bag of theories, and of little use to the practical man.

Such causes—namely, a general mistrust of so-called science, and to a small extent a painful experience in especial connection with fisheries, of the results of placing confidence in quacks who have falsely pretended to scientific knowledge—seem to me to be accountable for the fact that in the British Islands, neither publicly nor privately, has there been any attempt to make use of the services of scientific men in relation to our fisheries. The recent appointment of the distinguished naturalist who is at present Inspector of Salmon Fisheries,[1] is evidence of a new disposition to seek the aid of the highest authorities in science in connection with this subject; but it must be remembered that salmon fisheries form but a very small part of British fisheries in their entirety, and that a large staff of experienced naturalists would be required to deal satisfactorily, within a reasonable time, with the many important problems presented by the British Sea Fisheries.

The Governments of some foreign States, notably of France, but also on a smaller scale of Norway and Sweden, Holland, Prussia, Saxony, and in a special, and in many respects very noteworthy, manner, that of the United States of America, have concerned themselves to obtain the aid of zoologists in developing and managing the resources of the fish industries of their respective territories. The results of the application of accurate knowledge concerning fishes, and such shell-fish as oysters, mussels, pearl-mussels, lobsters and cray-fishes, have been in some cases strikingly successful; in other cases time has yet to show what advantages may result from the attempts which have been made. In all these countries, however, one very distinct result of the appreciation of the possible value of scientific knowledge of fishes and shell-fish by the

---

[1] Professor Huxley who, however, was not succeeded on his retirement by a scientific man.

State authorities has been this, namely, that zoologists are occupying themselves independently, and with increasing earnestness, with the investigation of all that relates to the life and growth, the food and the enemies, of the marine and freshwater organisms which form the material basis of fisheries.

In the present Exhibition, accordingly, we see not a few of these scientific results exhibited in the courts assigned to foreign exhibitors; whilst, on the other hand, in the British department there is very little which comes under the head of zoological science at all, that is to say, which illustrates the results of exact inquiry into the natural history of the fishes and other animals which are such an immense source of wealth and industry to our seafaring population.

Before proceeding to enumerate and describe these scientific collections, I should wish briefly to explain in what ways it seems probable that the accurate knowledge with regard to fishes which is now being accumulated by zoologists may hereafter be useful in the regulation and management of fisheries.

In any given area of land or water, under natural conditions, where animals can obtain nourishment, there is found living (taking one year with another) fully as much animal life as can there nourish and reproduce itself. Practically the whole of the earth's surface and of the sea is fully taken up by plants and animals. Many thousands more of most kinds are annually born than can possibly survive to maturity. The number of each kind of animal in natural conditions does not increase; but there is a strict balance maintained, so that, with local exceptions here and there, those that survive to maturity in the struggle for existence merely replace those which have gone before. Many thousands of young perish, serving as nourishment for other lower and higher organisms, whilst the total number of mature organisms remains the same.

Take as an example the microcosm constituted by a pond in which carp are cultivated, as in Germany, where these fish are valued as food.[1] Such a pond is allowed to remain dry every fourth year, when it is cleansed and puddled. Water is then allowed to run in, and a given number of one-year-old carp are placed in it. After three years these are taken out and sold as food. In one such pond 30,000 young carp have been observed constantly to yield at the end of three years 20,000 kilogrammes of marketable carp; when more than 30,000 young carp have been placed in the pond, no greater yield has resulted, 20,000 kilos. weight of carp-flesh is the result of that pond's activity after three years. The food of the carp consists of delicate vegetable growths, of insects, and other minute aquatic animals. These hatch from eggs and spores introduced with the water into the pond, and the pond will only carry such an amount of this food as will in three seasons produce 20,000 kilos. of carp-flesh. It is further found necessary to keep with these carp in the same pond a few pike, which prey upon the carp to a certain extent. The carp-culturists know how many pike to introduce. A few act beneficially in

[1] I borrow this illustration from Professor Möbius of Kiel.

destroying the smaller and more sickly individuals of the carp-stock, and so prevent valuable vegetable and insect-food from being consumed by individuals who would either not survive for the three years, or would show no growth proportionate to their consumption of food. On the other hand, too large a number of pike would reduce the total weight of carp, and leave much of the minute food in the pond unconsumed, whilst a large portion of it would have been converted into pike-flesh instead of remaining as carp.

In the limited area of the carp-pond there are a great number of processes going on, which contribute to the ultimate production of the 20,000 kilos. of carp-flesh. The minute vegetable organisms are continually feeding on the carbonic acid absorbed by the water from the atmosphere, and on the nitrogen partly existing originally as nitrates and ammonia therein, partly returned to the water by the excretion and decay of its animal inhabitants. Minute worms and crustacea are feeding on these plants, and other larger insects are feeding on these; finally, the carp nourish themselves on all these living things, and are to some extent preyed upon by the pike. Definite physical conditions, such as the presence or absence of a stream in the pond, the extreme heat and cold of summer and winter, and the presence of saline constituents in the water, determine the excess or the absence of one or another of the lower forms. If only *one* of these conditions be varied, the whole balance may be upset. An excessive growth of some minute plant, such as an Oscillatoria, favoured by heat or by the destruction

of some other organism, may lead to the destruction
of the proper food of the carp, and the yield of the
pond may be endangered.   The whole of these circumstances can, in the case of the carp-pond, be studied
and controlled.

If we now pass to the consideration of any given
area of the sea-bottom, we find that, though the area
is not definitely limited, the same interaction of the
various organisms inhabiting it holds good.   One
form is preying upon another, and determining by its
presence the numbers and the interaction of all the
others.   Physical conditions which affect one form,
may in the same way, as in the carp-pond, affect the
prosperity and abundance of another form.   Currents,
varying seasons, and such-like conditions, must obviously produce their effect.   But still more influential
must be the operations of man in removing a large
number of edible fish from such an area.   It is a mistake to suppose that the whole ocean is practically
one vast storehouse, and that the place of the fish
removed on a particular fishing-ground is immediately
taken by some of the grand total of fish, which are so
numerous in comparison with man's depredations as
to make his operations in this respect insignificant.
Even were it proved that there is this sort of cosmo-
politan solidarity about such fish as the Cod, which
live in deep water, there is, on the contrary, evidence
that shoal fish, like Herrings, Mackerel, and Pilchard,
and ground-fish, such as Soles and other flat-fishes,
are really localised.   If man removes a large proportion of these fish from the areas which they inhabit

the natural balance is upset, and chiefly in so far as
the production of young fish is concerned.   It is true
that several thousand young are produced by each
pair of fish left in the breeding area, and it might be
argued that since, in the absence of man, only two
out of the many thousand born of each pair of fish
come to maturity and breed again in their turn, the
only result of man's depredations (in addition to the
depredations of other enemies) is to make way for
more of the young, and to enable more than two of
the many thousands born of each pair in a preceding
generation to survive and breed in their turn.   This
argument is at once seen to be fallacious when we
remember that the thousands of apparently superfluous young produced by fishes are not really superfluous,
but have a perfectly definite place in the complex
interaction of the living beings within their area.
These very young fish serve as food to other forms,
which in their turn are fed upon by others, and are so
interwoven with the necessities and conditions of life
of other inhabitants of the area, that to remove, say
something like a fifth or even a tenth of them by
removing the parent fish, must cause a serious disturbance in the vital balance of that area.   When
the fisherman removes a large proportion of soles
from a given area, and so reduces the number of
young soles born in the same season in that area, he
does not simultaneously destroy the natural enemies
of the young soles : consequently very nearly the
same number of young soles are destroyed by such
natural enemies as were so destroyed before man inter-

fered, although very many less young soles are produced. It is thus quite clear that there is in reality
no reserve stock of young to take the place of the
adults removed by the special interference of man.
The increasing scarcity of the sole is a serious fact,
and is thus to be explained.

From this point of view it is clearly important, if
we wish to keep up the number of food-fishes in an
area which is fished by man, or to increase that
number, that we should (1) either know what are the
natural enemies of the food-fish in question at various
stages of its growth, and seek to destroy those enemies
in proportion as we remove the adult fish; (2) or,
again, that we should isolate and protect the young
fish from these natural enemies for a part of their
lives; (3) or, lastly, that we should, in proportion as
we remove breeding fish from the area, artificially
introduce into that area eggs, or young fish hatched
under supervision, so as to supply the deficit created
by the fishery of egg-producing adults.

Any of these operations requires very considerable
and most accurate zoological knowledge, and it would
be madness to attempt to carry any of them out by
proceeding upon hasty guesses or suppositions as to
the habits and life-history of the animals concerned.

There is also no doubt that certain modes of fishing and seasons of fishing may be more destructive,
more disturbing to the balance of life in a given area,
than other modes and other seasons of fishing. The
food of the fish which are valued may be destroyed
by some of man's operations, their enemies may be

unwittingly encouraged by others. Legislation is continually demanded, and has been from time to time carried out, in reference to such matters as modes and seasons of fishing and pollution of waters. But it is undeniably true that, in most cases, the accurate knowledge as to the life-history and circumstances of fishes is too small to justify legislative interference. No doubt zoologists have suggested some valuable restrictions which have been adopted by the Legislature in regard to some fisheries, and it is to Linnæus, the great Swedish zoologist of the last century, that Sweden owes important fishery laws. But if we are to have effective legislation at the present day in regard to our sea fisheries—we must, before proceeding any farther, have *more knowledge*. Those (and there are many) who earnestly desire additional restrictive Fishery Laws should do their utmost to enable zoologists to carry on researches which will provide that accurate knowledge of fishes and shell-fish, their food-reproduction and conditions of life—which must be obtained before legislation can reasonably be proposed.

The only mode of deciding between the conflicting opinions which have so often been expressed during this Congress, as to the necessity of this or that legislative enactment, is by bringing new knowledge to bear upon the questions at issue. That new knowledge is nothing more nor less than a part of Zoological Science, and can only be obtained through the exertions of those who are already acquainted with the actual condition of that science, and with its methods of minute and thorough investigation.

It is apparent, then, that the results of zoological science, as they may possibly affect fisheries, must even at the present moment be very considerable. We may classify them as follows:—

I. The discrimination and classification of the different kinds of plants and animals, including the fishes themselves which inhabit the various fresh and sea waters where fisheries are carried on. This is what is known as the study of Systematic Zoology, and of the fauna and flora of districts.

II. The knowledge of the successive phases of development or growth from the egg, and of the internal anatomy and mechanism of life of the chief forms of such animals and plants. This constitutes what is known as General Morphology and Physiology.

III. A specially detailed knowledge of the life-history of those species of fishes, molluscs, and crustacea, which are valuable to man and are the subject of fisheries ; a knowledge of their migrations, susceptibility to external influences, of their food and its history in detail ; of their enemies, in the shape of other fishes, birds, whales, seals, and insects, etc., which prey upon them and their young ; a knowledge of their parasites, injurious or harmless, and of their diseases. Such knowledge may be termed the Special Biology of Economic Fishes.

IV. A knowledge of those particular features in the life-history of an economic fish or mollusc, which directly concern the work of the fisherman or the fish culturist ; a knowledge of the effects produced by

particular fishing operations, as shown by statistics,
or of the effects produced by particular methods of
preservation and culture : further, a special knowledge of those parts or qualities in economic fishes or
molluscs which are of commercial value, and a knowledge of methods of improving or securing those parts
or qualities. This group of topics constitutes what
may be called Pisciculture.

Coming under the first head—of Systematic
Zoology—there are some valuable collections in the
present. Exhibition, but on account of the large space
which they would occupy were they complete, such
collections are, on the whole, rather samples of larger
collections than attempts at complete illustration of
the marine or fresh-water inhabitants of a district.
Thus, Dr. Dohrn of Naples has sent a series of about
400 bottles containing specimens in a marvellous
state of preservation, accurately named, of the fishes,
crustacea, molluscs, annelids, star-fishes, corals and
jelly-fish of the Bay of Naples. Mr. Oscar Dickson
sends also a very beautiful collection of named specimens from the Gothenburg Museum, illustrating the
fauna of the neighbouring sea. Professor Lilljeborg
exhibits in the Swedish department a very large
scientifically named collection of the crustacea which
form the food of many fishes in the great fresh-water
lakes of Northern Europe. He has especially occupied himself with the study of these organisms, and
has discovered many new species; it is worthy of
remark that English naturalists only two years ago
became alive to the fact that the same fresh-water

crustacea exist in the English and Scotch lakes. It is quite possible that a proper knowledge of these crustacea may at some future day be of value in attempts to cultivate fish in British lakes. Marine birds and mammals are exhibited in various parts of the Exhibition. Dr. Francis Day exhibits his great collection of Indian Fishes—preserved in spirits, and accompanied by the coloured plates of his great book on Indian Ichthyology : he also exhibits a collection of British fishes carefully preserved and named. The American department is remarkable for the carefully coloured series of casts, representing different species of the American food-fishes, and for samples of the animals of lower classes obtained from considerable depths off the American coast. Complete collections of the edible crustacea and mollusca of the United States, and of the commercial sponges of the coast of Florida, are also exhibited.

Under the second head—viz. General Morphology and Physiology—there is very little to be noted in the Exhibition. In fact, when we have mentioned the series exhibiting the growth of the salmon from the egg onwards, exhibited by Professor M'Intosh of St. Andrews, and the series of flat fishes of various ages in Mr. Oscar Dickson's collection—there is nothing except the valuable drawings of the anatomy of the oyster, and of its development from the egg, exhibited by the Netherlands Society of Zoologists. Under the auspices of this Society, which possesses a movable house fitted as a zoological laboratory, which can be erected for temporary use on any part of the

Dutch coast—two Dutch naturalists, Dr. Hoek and Dr. Horst, have within the past two years made some careful studies of the oyster which have very greatly added to our knowledge of that important mollusc, and may eventually be of service to the oyster-culturist. The results obtained by these observers are shown by large coloured drawings exhibited in the Netherlands department.

Some results of zoological science coming under the third head, viz. Special Biology of Economic Fishes, are to be found scattered here and there in the Exhibition. Collections of insects injurious to fresh-water fish or to their eggs and young, are exhibited in the Swedish Court, also, in a small way, in the American Court. A few parasites of fishes are ex-hibited by Dr. Cobbold, in the Eastern Arcade, and a remarkable series of the crustacean parasites, or fish-lice, of the fishes of Trieste is shown by Dr. Antonio Valli in the Austro-Hungarian section. Fish diseases are represented by stuffed specimens of salmon, with cotton wool attached, indicating the position of the growths of Saprolegnia which cause the malady known as the 'salmon-disease.'

The fourth section into which we have divided the results of zoological science, as seen in the Exhi-bition—namely, that of the various developments of Pisciculture—is richly represented by the exhibits of English oyster-culturists and salmon farmers; but in the most interesting way in the American depart-ment, where the devices made use of for hatching the eggs of sea-fish are shown. I need hardly say that the artificial hatching of the eggs of sea-fish is a novelty, and distinctly a result of the application of scientific knowledge. Since the culture of marine and of fresh-water fishes has formed the subjects of special papers at this Congress, I will not venture to say anything further about it here beyond claiming it as a scientific result.

There appears to be no exhibit in the building relating to the Pearl fisheries, either marine or fresh-water; in relation to these we should anticipate that the application of scientific knowledge might yield some very definite results (in the way of pearl-production).

Sponge fisheries are represented by collections of sponges from Florida, from the Bahamas, and from the Greek Islands. In the collection from Florida is a specimen having a considerable scientific interest. It represents an attempt at sponge-culture. Some years since in the Adriatic—under the Austrian Government—Professor Oscar Schmidt made some experiments on the propagation of sponges by cuttings. It was found possible to cut a live sponge into pieces, and affix these pieces each to a separate slate by thread, when each piece would attach itself to its slate and continue to grow. In this way sponges can be transported from one area to another —but the total weight of sponge is not increased, for the pieces of the divided sponge only produce the same amount of new sponge as they would have done had they never been separated from one another.

It appears that sponge-cutting is being attempted

on the Florida coast; and in the American Court are
two sticks with sponge-cuttings growing upon them,
which have been artificially placed there. I am not
aware as to whether any valuable result has been
obtained by thus cutting the sponges; but it is
certain that they might be thus introduced into
artificial basins, and grown there, were the general
conditions in such basins favourable.

Coral fishery is represented by an exhibit from
Naples, and by three pieces of valuable red coral
from Japan which have been purchased for commercial purposes at a high price. The eminent
French zoologist, Lacaze Duthiers, under the direction of the French Government, made a very thorough
study of red coral, and obtained scientific results of
great importance, fitted to assist the coral fishermen
in the regulation of their fisheries and the culture of
coral; but these results are not in any way illustrated in the present Exhibition.

On the whole, it appears when one attempts to
enumerate the results of zoological science in relation
to fisheries—illustrated in the present Exhibition—
that there are but few results which are so illustrated
—in fact not nearly so many as one might have expected. But, on reflection, it will appear that it is
difficult to show, in the form of a tangible exhibit,
many of these results. They are for the most part to
be found in books, and in the memoirs and illustrative
drawings published by scientific societies. Microscopic
preparations, showing every detail of the growth from
the egg—of the oyster, the mussel, the lobster, the

sole, the cod, and salmon, are hardly to be looked for. Really complete zoological collections would be too cumbrous for transport and exposure in such an Exhibition. But above all it is true that many of the most important scientific conclusions affecting the interests of fisheries are not capable of exhibition. The instruments with which the investigations are made, and in some cases the animals which have been the subject of investigation, may be exhibited, but the scientific result can often only be "exhibited" in so far as it affects the procedure of fish-catchers, fish-breeders, or fish-culturists.

I think that it has been made very apparent, not only by the class of objects exhibited by foreign contributors to this Exhibition, but also by the original papers and the discussions which the Conferences connected with the Exhibition have produced, that there is nothing which is so much needed in connection with all kinds of fisheries, river or sea, shell-fish, true fish, coral or sponge—as more knowledge, more science—in fact, more zoology; and not only that, but that there is nothing which is more desired and recognised as needful by all those who are best informed in their own particular branches of fishery.

Improved machines for catching fish, new legislative restrictions, State aid to fisher-folk—all such desiderata are, I believe, admitted to be less urgently needed, less likely to prevent our various fisheries from deteriorating or disappearing altogether, than the one desideratum—more accurate knowledge.

It is admitted on all sides that many British

fisheries are suffering, or are in a precarious state—others are actually destroyed. It is also admitted that our only chance of bettering this state of things is an increase of scientific or accurate knowledge.

If this is the case, there will be one grand scientific result of the International Fisheries Exhibition, and that will be an increased attention to, and adequate provision for, the carrying on of zoological studies in relation to fishery-animals.

I can picture to myself the shape which this scientific result might take, and I should be very glad were it to commend itself to the many influential men connected with fisheries who have organised this Exhibition, and will have the direction of its final outcome.

If it is demanded that more accurate knowledge of fishery-animals shall be provided for the public use, then arrangements must be made to enable skilled zoologists to carry on the investigations required. To make such investigations, continuous residence for weeks or months at a time, by the sea-shore, is necessary. In France, Holland, Italy and the United States, sea-side laboratories have been constructed, which are provided with working-tables, glass apparatus, aquariums, etc., and a staff of attendants and fishermen—to which naturalists can resort who desire to carry on investigations upon the life-history of marine organisms. Very valuable researches have been made through the agency of these institutions, and there can be no question as to the facilities which they afford, and the inducement which their existence offers, to naturalists for occupying themselves with these particular studies.

By offering free accommodation in such a laboratory to competent investigators you may obtain a large amount of valuable results at a minimum of expenditure. In any such laboratory there would probably be one or two permanent officials who would be competent zoologists, charged with special subjects of investigation and receiving salaries—but in addition to these, the laboratory would throw open its resources to voluntary workers (as do the foreign laboratories of which I have spoken), and thus the working power and the general interest of the scientific world in these institutions and their work would be enormously increased.

I can imagine a National Fisheries Society or Association, such as may come into existence in connection with this Exhibition, building such a laboratory for the study of marine zoology in relation to fisheries, somewhere on the coast not too distant from London. Such a laboratory would stand near the shore, possess its own jetty and small harbour, with steam-launch for dredging and trawling, and other boats. Adjacent to it would be marine ponds for experiments in the culture of oysters, mussels, and whelks, and of various fish. The director of the laboratory and his assistant would be provided with houses forming part of the laboratory building. The basement of the laboratory would consist of large well-paved rooms fitted with tanks, and an apparatus for the circulation of sea-water. Here animals would be kept for observation, and the produce of a day's dredging or trawling would here be sifted and sorted. On the ground-floor and first-floor would be spacious rooms, with large windows giving both north and south lights, and fitted with tables suited to the requirements of the microscopist. Small aquariums and pumping apparatus would also be provided in these rooms. Accommodation for ten workers, in addition to the director and his assistant, would thus be provided. In another room a complete zoological and piscicultural library would be established, and the means for writing and making drawings would be provided.

The naturalists permanently and temporarily working here would in the course of a few years provide us with much-needed knowledge. For instance, some would study the reproduction of the sole, and devise means for increasing its numbers in the market ; others would ascertain how best to deal with oysters ; others would find out the whole history of the mussel. Bit by bit a new and thorough knowledge of fishery-animals would be built up, and come into use as the basis of new legislative enactments, and of new methods of capture and culture.

Such an institution would no doubt be costly. A valuable laboratory of the kind might be set going and carried on for a smaller sum ; but a really creditable and efficient laboratory of the kind, with a first-rate man of science for its director, would cost £15,000 or £20,000 to establish, and some £3000 a year to maintain.

These figures and the whole suggestion may

appear extravagant to some persons who are not aware of what has been done elsewhere.

The "Zoological Station" of Naples, founded and directed by Dr. Anton Dohrn, is no less costly and efficient an institution than that which I have briefly sketched. Elsewhere, at Trieste, at Concarneau, at Roscoff, at Beaufort in North Carolina, less costly institutions have been set going, which are doing most valuable work.

The British coast is entirely destitute of any such home of research. No zoologists are employed by the Government or other authorities in this country to investigate our fishery-animals: those zoologists who do carry on such work on the coasts of this country, do so at their own expense. There is not even a laboratory, a boat or a dredging apparatus, provided by any public body to assist them. Naturally enough their work has not been hitherto specially directed to problems connected with fisheries; but it is only needful to offer to the many isolated investigators the use of a good laboratory and a well-considered organisation, in order to obtain through their co-operation the new knowledge which is so urgently needed.

I cannot but think that it will be a matter for profound congratulation to those who have brought this Exhibition into existence, to the Legislature, to men of science and to those concerned in fisheries, should we be able in future to point to a "National Laboratory of Marine Zoology" as THE Scientific Result of the London International Fisheries Exhibition.

APPENDIX

THE hoped-for "scientific result of the Fisheries Exhibition" sketched at the end of the lecture given by me in 1883 has become a reality. At the close of the Exhibition I obtained the signatures of a number of leading naturalists and others interested in fishery questions to a memorial to the Executive Council of the Exhibition, asking that a portion of the surplus receipts of the Exhibition might be devoted to the foundation of a laboratory for the production of new knowledge with regard to food-fishes. The surplus was, however, applied to other purposes, and the chances of starting such a laboratory seemed small, when Dr. Günther of the British Museum suggested to me the formation of an association having for its object the foundation and maintenance of one or more such laboratories. Some of the officers and Council of the Royal Society warmly supported this suggestion, and kindly gave the use of the rooms of the Society in March 1884 for a meeting at which Professor Huxley presided. This meeting was attended not only by scientific biologists, but by representatives of the Fishmongers' Company, and by all the members of the Royal Commission on Trawling, at that time engaged in inquiries into the fishing industry. It was agreed to establish the Marine

Biological Association of the United Kingdom, of which Professor Huxley became president, and H.R.H. the Prince of Wales patron. About 300 members joined the Association, and subscriptions were invited for the purpose of erecting a laboratory. After due inquiry Plymouth Sound was agreed upon as the locality for the first laboratory, and a site was placed at the disposal of the Association by the authorities of the War Office. Donations of £500 and upwards were received from the Fishmongers' Company (£2000), the Clothworkers' Company, the University of Oxford, the University of Cambridge, the British Association for the Advancement of Science, Mr. Robert Bayly, and Mr. John Bayly of Plymouth. Sums varying from £100 to £300 were subscribed by the Drapers, Mercers, Goldsmiths, and Grocers' Companies, by the Corporation of London, and by the Royal, the Zoological, and the Royal Microscopical Societies, as well as by a number of gentlemen interested in the enterprise, either as naturalists or as philanthropists. Finally Parliament, on the recommendation of the Treasury, voted a sum of £5000 in aid of the building and fitting of the laboratory, and a sum of £500 a year for five years towards defraying the annual cost of management and research. The laboratory, which is represented in the vignette, stands in front of King Charles's Curtain on the Citadel Hill, overlooking Plymouth Sound. It was completed and opened for work in June 1888. The building and fittings, including aquariums, pumps, and steam-engines, have cost about £12,000.

It is at present carried on at an annual expenditure of about £1250. Its staff includes Mr. Gilbert Bourne, M.A. Oxon., director; Mr. Garstang, B.A. Oxon., assistant; and Mr. Cunningham, M.A. Oxon., naturalist. It possesses a fine library, a steam launch, a yawl, and a row-boat, and employs a fisherman, engineer, laboratory servant, and other hands. A large number of naturalists have availed themselves

PLYMOUTH LABORATORY

of the facilities afforded by the laboratory, which contains ample accommodation for fourteen workers, each occupying a private work-table, screened off from neighbouring tables, and fitted with fresh and salt water, gas, chemical reagents, etc. The Association has published four numbers of a journal containing reports of the work done at Plymouth, as well as the official reports of the Council. The Council meets in London once a month to superintend the finances of the Association, and direct in a general way the work carried on at Plymouth. From the foundation of the Association I have had the pleasure of acting as its honorary secretary, in co-operation with a Council consisting chiefly of scientific naturalists, who have given time and trouble freely for the purpose of establishing the Plymouth Laboratory as a national institution. Already the Association has been the means of adding important facts to our knowledge of that most valuable food-fish the common sole—a large and richly-illustrated volume by the naturalist specially employed for these researches—Mr. Cunningham —being at this moment in the press. The anchovy, the pilchard, the lobster, and the oyster are each receiving special attention, which will in due time lead to practical results. Besides these, a large number of researches on purely scientific questions, such as the electric organs of skates, the anatomy of Crustacea, the structure of new and rare minute organisms taken in the tow-net or in the pools at low tide, have been carried on by naturalists who have come to Plymouth for the purpose of profiting during a month or two by the facilities offered in the laboratory of the Association.

The task which now lies before those who have founded this institution is by no means a light one. It has to be carried on from year to year, and funds have to be obtained for this purpose. Those who may be interested in the publications of the Association, or desire further information concerning it, should apply by letter to the Director of the Laboratory, Citadel Hill, Plymouth.

VI

CENTENARIANISM

From *Macmillan's Magazine*, October 1871

.

CENTENARIANISM

THE somewhat unwieldy word standing at the head of this page is coined in order to take the place of that much-abused term "longevity," which is often made to do duty in a restricted sense, to its detriment. Longevity simply means "length of life"; and it can serve no good purpose to limit its application to those cases of length of life which are beyond the normal period among men : it is required for more general use ; and hence we may, with advantage, speak of old people who reach or exceed one hundred years of age as examples of centenarianism, instead of calling them examples of longevity. Every now and then, with more frequency and regularity than is presented by perhaps any other periodic topic, centenarianism excites the public interest. Another case is announced of an individual having exceeded one hundred years of age : paragraphs go the round of the newspapers, the medical journals report on the case, Sir George Cornewall Lewis is declared to be refuted, and the subject drops. It is a little strange at first sight, this interest which is manifested in monstrosities of life-duration. The men and women who have so far distinguished themselves among their

Q

fellow-creatures as to exceed greatly the average
height, have never attracted so much attention as have
the long-livers; and yet it is probably as rare for a
man to exceed eight feet in height as to live beyond
the hundredth year,—indeed, we believe much rarer.
No one asks the details of the life of an eight-foot
giant—how much pudding he took as a boy in order
to attain his astounding dimensions—apparently
because nobody believes that any administration of
pudding or its correlatives would make a boy, who
was *going to be* five foot four, into a man of larger
size. Possibly, moreover, not very many persons are
greatly anxious to attain large dimensions. It is not
so, however, with long-livers: even to-day all classes
of society take an interest, which is sometimes profound, in the details of life of a long-liver; they
would fain imitate the centenarian, and by copying
his mode of living inherit his years. Even where
there is no intention of pursuing a system of diet and
manner of life, people seem to like to know how they
could, if they chose, lengthen their years. There is
a relic of the old times, of the search for the *elixir
vitæ*, in this kind of thing: that great enthusiasm of
past days, which served an important part in opening
for us the door of science, is still alive. Clearly the
people who take more interest in the lesson to be
learned from the diary of a centenarian than from
the report of a Registrar-General or a medical officer
of health, are yet mediæval in their views of life and
death. The real fact seems to be that the man who
exceeds one hundred years of life has no more to

teach us than the man who exceeds eight feet in height : both are monstrosities, and attain their special distinction by no particular behaviour on their part. A certain amount of care will produce its due effect on the longevity of any individual; but there is a set limit beyond which it cannot be extended. In some individuals this limit is at a greater distance than it is in the most of mankind, and if they escape the accidents of disease and violence they live longer than other men : the cases of these men must be looked upon as distinctly abnormal; they are to be held as freaks of nature, monsters—giants of age; just as we have converse cases recorded of dwarfs of age—human beings who became old after twelve years of life, and began to exhibit senile decay at a time when ordinary men are still growing children.

Longevity, as we have elsewhere pointed out,[1] is of several kinds, which need to be distinguished. There is the longevity characteristic of species of plants and animals, men included,—that is to say, the age which each individual of the species born may be expected to reach ; this is *average specific longevity*, and is a very low figure indeed as compared with other kinds of longevity. For Europeans it does not appear to be above forty years. This average longevity is brought to so low a figure by the great amount of death in the first years of life. By an excess of deaths in early life the average longevity of a species or of any given group of individuals might be brought down to a year or two, though the individuals which

---

[1] *Comparative Longevity in Man and Animals.* Macmillan, 1870.

did survive might, some of them, enjoy a century of
life. This brings us to a second kind of longevity
also characterising species—that which agrees with
what has been called "the lease of life," and which
we call *potential specific longevity*. The age to
which a creature would attain, supposing it to escape
all the dangers of youth, the diseases and accidents
which are lurking about the life-way, and to die
simply of old age, would represent the "potential
longevity" of that kind of plant or animal. Very
few beings ever manage to exhibit this—certainly
very few men; but men are sufficiently anxious about
the matter, and many have taken so much pains to
live long, by avoiding all dangers, that we have good
ground to suppose that the lease of life of the present
race of men is normally something between seventy
and one hundred years. Care may enable a man to
expend very nearly his full lease; but nothing which
he can do, no power under heaven, can enable him to
add a day to that term, any more than by taking
thought a cubit may be added to his stature. And
now we see the relations which centenarians hold to
other men in this matter. They are not persons who
have taken more care than the less rare but equally
admirable octogenarians; they have simply been
born with a greater potential longevity—a longer
lease of life—and they have had the good or bad luck
to remain tenants for very nearly as long as the lease
was good. It is impossible to guess how many, but
doubtless thousands of possible centenarians die
before they are a year old, and thousands more at all

ages : had they managed to pass the one fatal corner where they fell, the whole road would have been clear for a hundred years.

Regarding then, as we do, centenarians as instances of extreme or "abnormal longevity," of which it is worth remarking we have two forms, the abnormally small[1] and the abnormally great, we can see no reason for fixing the limit of the abnormally great at one hundred years, as Sir George Cornewall Lewis was at one time inclined to do, nor even at one hundred and three or four, to which limit he was afterwards induced to advance. Our *à priori* impressions are distinctly in favour of a much wider limit, reaching perhaps, in the very rarest cases, to the one hundred and fifty years attributed to some celebrities, such as Old Parr, Henry Jenkins, and the Countesses of Desmond and Eccleston. Indeed, the great German, Haller, has uttered what is probably the truest dictum yet put forward in the matter : " The ultimate limit of human life does not exceed two centuries : to fix the exact number of years is exceedingly difficult."

When an unusually well-attested case of centenarianism turns up—as, for instance, the recent one of Mr. Luning, at Morden College, Blackheath,—the newspapers and journals always bring in the late Sir George Cornewall Lewis, attribute certain opinions to him, and demolish them by aid of the new case.

---

[1] An instance was not long since recorded in one of the medical journals of a child which ceased to grow and commenced to exhibit signs of senile decay at the age of ten years.

This is one way of keeping up the interest in the specimens of abnormal longevity; but inasmuch as several well-attested cases of persons exceeding a hundred years of age were adduced at the time when Sir George was interested in this matter, and were actually admitted by him not long before he died as sufficiently conclusive to make him modify the opinion he had held, viz. that there was no proof of the existence of centenarians, we are fully warranted in concluding that the importance attached to such cases from this point of view is as delusive as is the interest they gain from the supposition that we can learn by them how to live long ourselves. What Sir George Lewis at one time stated (it was during the last few months of his life that he brought his valuable sceptical criticism to bear on the matter) was, that he could find no sufficient proof of any man or woman having exceeded, or even completed a century of life; and having found so many cases advanced on the slenderest and most worthless evidence, he was inclined to regard all centenarianism as either delusion or imposture. In this he reminds us of a remark made by Professor Huxley: "No mistake is so commonly made by clever people as that of assuming a cause to be bad because the arguments of its supporters are to a great extent nonsensical." Sir George fell into this error, as he afterwards had to acknowledge; for upon the evidence which the publication of his incredulity brought down upon him in abundance, he was compelled to admit that persons do reach one hundred years of age, and that some

have attained even one hundred and three or four,
though this he considered exceedingly rare and as the
ultimate term of life.

By far the larger number of cases of centenarianism
which are reported are not backed up as they should
be by evidence. The appetite for the marvellous is
so keen, that people would rather take the centenarian
on his own assertion than risk losing him by investigation. This is the case with a certain Thomas
Geeran, now receiving parish relief at Brighton, who
is declared to be one hundred and four years old, and
states that he entered the British army at thirty years
of age, and served for more than thirty years. A
pamphlet has been published concerning this case, in
which there is not a shred of evidence given in support
of the man's statement. No inquiries appear to have
been made at his reputed birthplace, viz. Scariff, county
Clare, Ireland, and an application to the War Office,
with a view to getting him a pension, has entirely
failed, in consequence of his name not being discover-
able in the books. This is the kind of case which we
must guard against, and others like it, testified only
by epitaphs or village gossips. The next generation
will not be troubled with this question as we are to-
day, for the registration of births will, in the course
of time, furnish all the required evidence on one
point, whilst the only remaining difficulty, that of
establishing identity, is daily decreasing with the
growth of intelligence and the spread of education
among our peasantry.

It is to be hoped, however, that we shall not have

to wait so long for journalists and enthusiasts to cease their triumphant paragraphs, announcing cases in which the age of one hundred or one hundred and four years has been attained. Anything over one hundred and nine, in the way of age, would be perhaps worth mention if accompanied by documentary evidence; but of the mere passing the century limit there is enough proof already.[1]

We shall here briefly mention five cases of centenarianism, of the thorough trustworthiness of which we feel no doubt; and were it worth while, we fully believe that a great many others could be placed on an equally sure basis. The trouble and worry of doing this kind of thing is, however, not at all inviting; and where so little is to be gained either in the way of knowledge or amusement, we do not wonder that published well-attested cases are fewer than they might be.

1. William Shuldham was baptized at Beccles, in Suffolk, in July 1743. He died in May 1845. His baptism is witnessed by the register in the parish church of Beccles. On 22d July 1843 he gave a dinner at Marlesford Hall, near Wickham Market, to his friends, to celebrate the completion of his hundredth year.

2. A Quaker gentleman, well known in the mercantile world at the beginning of this century, died not long since in his hundred and second year. Dr.

---

[1] A great number of cases of centenarianism—good and bad—are given by Mr. Tollemache in an excellent article in the *Fortnightly Review*, April 1869.

Dickinson, of Mayfair, who has been kind enough to inform me of this case, has copies of the register both of his birth and death, establishing this fact. As Dr. Dickinson observes, the Quakers are very precise in these matters.

3. James Hastings, for upwards of sixty years rector and impropriator of the living of Martley in Worcestershire, father of Admiral Sir Thomas Hastings, Sir Charles Hastings, Admiral Hastings, and the Rev. Henry Hastings, died in his hundred-and-first year. His grandson, Mr. G. W. Hastings, of Barnard's Green, Malvern, has obliged me with the following details. He was born in London, in Soho Square, 2d January 1756; and his birth register, of which Mr. Hastings has a copy, is at St. Martin's, Trafalgar Square. He was entered as a gentleman commoner of Wadham College, Oxford, in 1776. At the request of Sir Thomas and Mr. Hastings, the warden of Wadham last year looked up the entries in the college and university books, and sent a copy of an entry, giving the age of James Hastings as twenty at matriculation. He was admitted to holy orders by the Bishop of Oxford, at St. Mary's, in November 1779. As no one can be admitted to the orders of the Church of England till the age of twenty-three, this again carries him back to 1756 as his birth-year. Mr. Hastings has the letters of orders in his possession; they have never left the family, and prove incontestably that James Hastings was twenty-three in 1779. He was married at the parish church of Chipping Norton in February 1781, and his age is

given in the register as twenty-five. He died in July
1856, and was buried in the family vault in Martley
Church. The Rev. James Hastings stood six feet four
inches in his stockings, was a strikingly handsome
man, and had fifteen children. He had but one
sister, and no brother, whilst his wife had one brother
and no sister. His father did not much exceed sixty
years in age ; and Mr. Hastings informs me from his
family records, which extend to the time of Henry II,
that there are no remarkable cases of great age among
his earlier progenitors.

4. Captain Lahrbush in March 1870 celebrated in
New York city his hundred and fourth birthday anni-
versary. He was born in London, on the 9th of
March 1766. He entered the British army on the
17th of October 1789, and documents connected with
this entry prove his age at that time to have been
twenty-three years.

5. Jacob William Luning died recently, at Morden
College, Blackheath, in his hundred and fourth year.
Documentary evidence sufficient to satisfy Dr. Farr,
of the Registrar-General's Office, has been adduced,
proving that he was born at Hamelvörden in 1767,
and similar evidence of the date and the age he gave
when he was naturalised as a British subject, also
when he was married, and, what is still more important, when he insured his life—an occasion on which
men are not likely to add anything to their age.

As to the means by which to live long, and to
give ourselves the chance of enduring to our hundred
and tenth year, if we have it in us, or to our eightieth

only, if that be the limit of our "matter of life," we must consult the statistics which are available, and not try to draw any conclusions from these extreme cases. What will lengthen and what shorten life, however, becomes a question of general longevity, and this we did not propose to ourselves to discuss on the present occasion. We may, nevertheless, notice that everything seems to show that the appliances of civilised life, and quiet and regular habits, are the chief conditions of long life. Europeans are, it seems, longer lived than other men; and Englishmen than French, Germans, Swedes, or Belgians, as far as statistics tell us. In Lord Bacon's time there was a prejudice in favour of the wild Irishmen—"Hiberni sylvestres," as he calls them, who were in the habit of smelling the fresh earth and drinking infusions of saffron. Statistics and Saxon domination have deprived Ireland of this pre-eminence in longevity. We also find from statistics, comparing the expectation of life at the age of sixty, given by various authorities, that in England agricultural labourers of that age, belonging to friendly societies, and hence sober, well-to-do men, stand first, and may expect to live nearly eighteen years longer, whilst confirmed drunkards stand last, with only half that chance of life. The females of the aristocracy come next to the labourers, with sixteen years and a half; the male members of the aristocracy next, with only fourteen and a half; clerks follow, with twelve and a half; men in Liverpool, with twelve; miners, with eleven and three-quarters; whilst sovereigns of all countries at sixty years of age have an expectation of a little less than eleven years of life. Distinguished men live a shorter time than less distinguished, on account of their harder work; married live longer than un-married persons, on account, perhaps, of the measured tranquillity of connubial life; women longer than men, because they lead an easier life; and the clergy longer than other professional men, for the same reason.

From these facts it is not difficult to draw the lesson of longevity. After all, the prolonging of their own lives is not a thing about which men should take much thought; as long as they are careful not directly to shorten life, and careful to preserve health, longevity and centenarianism may well be left to take their own way. The celebrated Italian, Louis Cornaro, carefully weighing his egg, and measuring his wine for his daily meals, refusing to allow matters of a disturbing nature to come under his attention, and taking a thousand precautions, all to enable his pitiful old frame to vegetate a few years the longer on the earth's face, is not a pleasing figure to contemplate. True it is, that he who would save his life shall lose it; for the existence of such a being as Cornaro is not comparable day for day with that of an active man. When the element of intensity is taken into considera-tion, there is perhaps very much less difference be-tween the quantities lived by various men than would appear from the simple record of time. But whilst it is not for the men of to-day to cherish the search for elixirs of life, nor to desire nor endeavour to become centenarians, there is yet a longevity which they can most materially influence—which they can check or extend by deliberate acts most directly, having it in their power to add years, hundreds of years, of life to the community—of active, vigorous life, too, not such as the common seeker of longevity would gain ; and this longevity it is no less our interest than our duty to work for. Men can diminish the mortality of populations by attention to simple laws of health, and, by increasing the average longevity, give that increased happiness and prosperity which security of life and health brings. It is in sanitary action that the *elixir vitæ* has been discovered in these days, which, though it perhaps has not as yet increased the roll of centenarians, has no limit to its operations, until the time shall have come when man will no longer, as Buffon said, " die of disappointment," but " attain everywhere a hundred years."

LIBRARY OF THE UNIVERSITY OF CALIFORNIA

VII

PARTHENOGENESIS

A REVIEW OF VON SIEBOLD'S *BEITRAGE ZUR PARTHENOGENESIS DER ARTHROPODEN*, FROM *NATURE*, 10th OCTOBER 1872

PARTHENOGENESIS

AMIDST the all-absorbing discussion of the problems which have arisen out of the general acceptance among biologists of the law of evolution, the phenomenon of Parthenogenesis which, previously to Mr. Darwin's work on the Origin of Species, excited the interest and called forth the investigations of observers in much the same manner as his theory has done of late years, has met with a reverse of fortune and fallen into a subordinate rank of popularity. The distinguished naturalist, however, who fifteen years ago gave so stunning a blow to current theories of the reproductive process, by demonstrating the occurrence in moths and bees of what he designated as "true parthenogenesis"—that is to say, the development, without impregnation, of an ovum capable of being impregnated—has not let the subject drop. Professor Siebold has made further experimental researches, establishing again, and on a larger basis, his former conclusion, and adding at least one new fact of great general importance for the understanding of the process of sexual reproduction. Although upon its first appearance in 1856, the conclusion arrived at in his *Wahre Parthenogenesis* was admitted by almost

R

all competent naturalists to be thoroughly demonstrated, and beyond the reach of criticism, yet some
more and some less eminent biologists have not been
wanting to deny the *Lucina sine concubitu*, and have
raised objections, such as that there is a possibility
of error in the experiments in reference to the exclusion of males from the supposed parthenogenetic
female; and again, that these so-called females were
not demonstrated "not to be hermaphrodites." Indeed so deeply rooted is the conviction that eggs are
made to be impregnated by spermatozoa, and that
they then, and then only, can proceed to develop,
that Siebold felt it necessary to add to his proofs,
in order to establish his position that not only do
unimpregnated eggs develop into perfect animals, but
that such an event is by no means an exceptional
occurrence among certain groups, and has a definitely
fixed and orderly recurrence amongst them. He
naturally was also anxious to extend the class-limits
within which a true parthenogenesis can be said to
occur, and he desired to inquire into the sex of the
parthenogenetically-produced offspring in such cases
as could be critically and decisively studied. Hence
the renewed researches which have extended over
several years, the results of which are given in the
present brochure.

Von Siebold's merit in this and his former work
(but more especially in this) is not the enunciation of
a new theory, or hypothesis, but the great care,
ingenuity, and persistence which he has displayed in
investigating cases in which for many years collectors,

bee-keepers, and such naturalists of the limited, or
" gardener" type, had asserted reproduction by means
of unfertilised eggs to take place. It must be remembered that he was himself a strong opponent
in 1850 of the supposition which he has now shown to
be justified in fact, and that Leuckart in his article
"Zeugniss," in Wagner's *Handwörterbuch*, and in
other publications, preceded him as an advocate for
the existence of true parthenogenesis, endeavouring,
by microscopical researches, to give a solid observa-
tional basis to Dzierzon's hypothesis. It was not
until 1857 that Siebold published his observations
on bees, demonstrating what had been previously
supposed, viz. that the queen-bee exhausts her store
of received sperm in fertilising eggs which give rise to
females only, and that then she lays unfertilised eggs,
which become drones only, whilst the unfertilised
worker-females also lay eggs which give rise to drones :
and again that in certain moths (*Psyche* and *Solenobia*)
unfertilised ova develop and produce females only.
Leuckart followed (1858) with his *Zur Kenntniss des
Generations-wechsels und der Parthenogenesis bei den
Insekten*. In this work, whilst asserting his claims to
the merit of first espousing the cause of true parthenogenesis, Leuckart gives an excellent view of the
general signification of the phenomena, and insists on
the importance of extended histological observation
in the examination of alleged cases of parthenogenesis.
In his present work Siebold cannot be charged with in
any way neglecting this part of his subject, for he has
given most important and minute descriptions of the

generative organs of the two principal cases studied
(*Polistes* and *Apus*), containing new facts. His
method is however eminently experimental, and
appears to us a striking contradiction of a very
superficial classification of the sciences, which is
favoured sometimes by men of science unacquainted
with the methods or problems of biology : we mean
the division into the exact or mathematical, the
experimental, and the classificatory sciences, in which
last division the so-called natural history sciences are
said to find their place.

The experiments which Siebold made on bees and
wasps, though performed by a naturalist, are as nicely
controlled, and as clear in the conclusions which they
give, as any performed by exact physicists on the
times or quantities concerned in this or that physical
process. The style in which details of these investigations are communicated is one rare at the present day
in biological works, where minute description of
structure, or of the apparatus devised for a physiological research, form the staple. Here we are treated to
a leisurely narrative of some years of patient work ;
we share the keen enjoyment of the author as he
becomes acquainted with the marvellous intelligence
of his wasps and their various proceedings—we feel
his satisfaction in overcoming the difficulties of pro-
curing and observing the necessary material, and
admire the candour and thoroughness with which
he handles the question before him.

Before proceeding to a short notice of the contents
of Von Siebold's book, it will be well to give a brief

statement of the signification which such inquiries as
his have in the present state of knowledge. Harvey's
dictum, " Omne vivum ex ovo," expressed a great law,
which had to be qualified when the researches of
Trembley and others made known, among Polyps, and
Worms, and Protozoa, reproduction by fission. To
this rapidly succeeded the recognition of a modified
fission, in which the animal did not divide into equal
parts, nor exhibit the power of reproduction of the
whole animal in artificially detached portions of its
body ; but in which special sprouts or buds were found
to be prepared and detached spontaneously, becoming
then developed into perfect animals. This process
received the name of gemmation, and was stated to
occur in polyps and also in the plant-lice. Parallels
for these methods of reproduction in animals were
readily recognised in plants, in the multiplication
by seed, by cuttings or shoots, and by separable
buds. A broad line was drawn between " buds " and
" eggs," however egg-like the former might appear, in
the assumption that eggs were special bodies of a
peculiar structure, destined to be " fertilised " by the
spermatozoa of the male—after which process only
could they develop. These distinctions, some twenty
years ago, were the more firmly impressed in the minds
of biologists by the then recently acquired knowledge
of the process of fertilisation or impregnation. Then
came the demonstration by Siebold of the capacity for
development of true eggs, even when not impregnated.
The sharpness of the limit between buds and eggs was
by this at once destroyed ; and the closely following

researches of Leydig (antecedent to Siebold's work in
some cases), Huxley, Lubbock, and Leuckart, on the
structure of the supposed buds of Aphis and allied
insects, and of lower crustaceans, proved that these
bodies were morphologically ova — originating in
ovaries, and having the essential structure of fertilisable
ova. For them the term "pseudova" was introduced
by Professor Huxley, since they differ in this respect
from other ova—that whereas the latter can be, and
are in most cases (though with constant exceptions),
fertilised, the latter cannot be.[1] Whilst, then, up to
this period such a thing as parthenogenesis appeared
to be a strange exception, the question has now shifted,
and, since the essential identity in reproductive power
of cuttings, buds, pseudova, and eggs is proved, the
problem before naturalists is rather "Why are eggs
ever fertilised?" in short, "What is the use of the
male sex at all?" We have animals and plants multiplying by fission, breaking up into two or more parts,
each of which becomes a new individual; we have
them giving rise by growth to masses of cells, which
become detached or remain attached, and develop each
into a new individual; and finally, we have them ela-
borating single large cells, which become detached and

[1] Leuckart has more recently proposed, in describing the reproduction of the Cecidomyia larvæ discovered by Wagner, to limit the term
"pseudovum" to such ova as those produced by larvæ, or imperfect
forms; and not to apply it at all to the eggs of bees, wasps, etc. (which
can develop without fertilisation), as was done by Huxley. The falsity
implied in the prefix seems to make a rather stronger distinction than
is desirable between any of these bodies; for they are all truly ova,
though ova of various special properties.

develop each into a new individual. Why should it be that in certain cases these last require fusion with another peculiar kind of cell elements before they will develop? Some light seemed to be thrown on this matter at first, by the observation that the unfertilised ova of the bee always produce drones, and that only the fertilised produce females; but this indication of a possibly clearer insight into the matter is entirely upset by the fact, now fully established in the present work, that in some species of insects and crustaceans the unfertilised ova always, or in an enormously large proportion, produce females only; whilst in the Aphides we know that they ultimately produce both males and females. Mr. Darwin has suggested the most satisfactory theory of fertilisation, in assigning to it the object of fusing two life-experiences in the progeny, which thus gains tendencies and acquires impulses from a wider area than does an unfertilised ovum, and is in so far strengthened. Conjugation of two cells, similarly formed but belonging to different individuals (as seen in Confervæ and Gregarinæ) is the simplest arrangement for obtaining this end; the only difference between this and sexual reproduction is that in the latter process one cell seeks, the other is sought; and this differentiation into active and passive, the wooer and the wooed, commencing in the simplest vegetable and animal cells, persists to the highest rank of development. Self-impregnation (if it have a real physiological existence) and parthenogenesis, have, then, to yield, as chief modes of reproduction, to gamogenesis, or the concurrence of two individuals; and this for one and the same reason. Perhaps the apparently anomalous facts that an animal—or plant, as the case may be—develops elaborate motile zoosperms and copulatory organs, merely to fertilise its own egg ; and that other animals and plants develop peculiarly constructed large cells, of a kind apparently especially elaborated in the progress of the general evolution of life to provide for fertilisation, yet which never are fertilised—are only to be explained as cases of persistent structures with modified function. In the former case, Agamogenesis, being sufficient to or necessitated by the conditions of life, yet avails itself of the apparatus inherited from gamogenetic ancestors ; whilst similarly, in the second case (pseudova), Agamogenesis, having advantages for the particular case (and not being a common phenomenon in the group), instead of making its appearance through new organs, avails itself of the ovary inherited from gamogenetic progenitors. Thus the unsignificant form of an ovum (unsignificant, that is, so far as Agamogenesis is concerned) takes the place of the more obviously appropriate bud or fission-product. The phenomenon of Alternation of Generations, usually treated of in connection with parthenogenesis, should by experiment on the physical conditions accompanying its variations enable us to ascertain a great deal more than is at present known of what is the signification of the differentiation of male and female sexual elements; and it is from further study of this and of true Parthenogenesis that progress in this part of physiology may be expected.

To return to Siebold's researches. The greater

part of the book is devoted to au account of the parthenogenesis observable in the wasp Polistes. Leuckart first recorded, in his work already mentioned, that the workers of wasps, ants, hornets, and humble bees lay eggs, which in one case he followed to the development of a larva, of which he was not able to determine the sex. Siebold determined to study a species of Polistes common in and around Munich, which he identifies with much care, and after reference to specimens and authorities from many lands, as *Polistes gallica* var. *diadema* Latreille. He gives a minute description of the characters of the females and males; the two kinds of the former (queens and workers) being only distinguishable by size — the workers in all external characters as well as in their generative organs being merely smaller queens, and fully capable of copulation and impregnation. In the beginning of May, in Munich, the Polistes queens which were born in the previous summer and impregnated then, commence each to build a nest. No queen who built in the former year survives to build a second time, and the young queens never make use of the old nests. The Polistes are very particular in choosing a warm sunshiny spot, sheltered from wind and rain, and as such spots are not too common, a new nest is often begun near the weathered remains of an old one. Walls and trunks of trees, often at such a height as not to be easily reached, are the sites chosen. When the queen has constructed fifteen or twenty cells, she lays eggs in them, and is very hard-worked in guarding her nest and in providing food for the larvæ as they hatch. She feeds them on caterpillars and other soft insects, first removing the alimentary canal (as cooks take out the entrails of a fish), and carefully applying the morsel to the lips of each larva. This process takes some time for one "hand," and hence the first brood is longer in coming to perfection than later broods, in the rearing of which the elder progeny assist. In the middle of June the first lot make their appearance, all small females ready to assist their parent in the advancement of her enterprise. The later broods develop more rapidly and acquire larger size from being better nourished, and towards the end of June (no drones being as yet born) the females which come forth are as large as the old queen ; they may, however, be easily distinguished from her by their comparative freshness of colour and wing. Great care is displayed in guarding the nest. At night the queen goes to sleep after having carefully inspected each cell, taking up her position at the hinder side of the nest. If she is disturbed in the night, she takes another survey of her house before again going to sleep. In the day-time if disturbed she makes an immediate attack on the enemy with her sting, and then flies back to her nest. She can sting several times, since the barbs on her weapon are not too long to allow of its being with-drawn. Ants are amongst the most common of the many insect enemies of these wasps, and when one ventures into the nest, the whole colony sting it to death, and immediately throw the body out. Birds are, however, not thus to be got rid of, and destroy im-mense numbers of the nests, so that Siebold was

obliged to protect those he wished to study with
nets. The members of one nest are not allowed to
remain in another, if by chance a stranger comes in
she is hustled out at once by the wasps near the entrance. Siebold convinced himself of this by painting
the thorax of a number of Polistes belonging to different
nests with different colours. Only late in the year,
when the wasps seem to be getting careless or tired of
their incessant work did he find that one or two had
got mixed in certain colonies, to which they did not
rightly belong. Although there is this sharp discrimination of individuals, yet it was found that by substituting one nest for another whilst the queen was
away, she could often be deceived, so as to make her
enter upon the possession of the substituted nest as
though it were her own. Siebold found this a very
useful plan when he wanted to change the position or
locality of a nest so as to bring it into a safer or more
accessible spot, or when a nest which he had been observing was by some accident deserted, or when a
nest in a favourable position was less forward in the
development of its larvæ than one less favourably
situated. By making the nests movable, and substituting the one for the other in the absence of the
queen, he was able to save himself much trouble and
loss. The nests were made movable by removing
them from their original support and firmly fixing
them to boards which were then hung up in the original
position. The queens were very anxious after this
operation had been performed, investigating with
great care the strength of the support and the cord by

which the board was hung, and sometimes adding to
it themselves additional strength. By degrees such
movable nests could be lowered a little bit each day
from an inconveniently high position, or taking the
nest in the night under a cover whilst all the wasps
were in it, it could be removed from a distant locality
to the Professor's garden. In such cases a certain proportion always came to grief by the desertion of the
colony; and the queen was then sometimes found at
work on the old site constructing a new nest. Although strangers are not admitted in a well-regulated
Polistes nest, yet by carelessness or desertion the brood
of one colony will sometimes be exposed to the attacks
of the workers of another, who then make use of the
unfortunate larvæ to feed their own young. It frequently happens that workers who have once indulged
in this kind of thing, become what are called "robber
wasps," utterly demoralised, and actually undo the
whole labour of a colony by dragging out the grubs
which they were lately so carefully tending in common
with their fellows, to feed the still younger larvæ.
When this condition of things has once begun in a
colony it soon goes to ruin, and hence it is necessary
to destroy any deserted Polistes nests in the neighbourhood of those under observation, lest by entering the
former the members of the latter should get the bad
habit of pulling grubs out of their cells, and proceed
to do the same in their own nests.

The rain is a very constant source of destruction to
Polistes colonies, drowning the young by saturating
the cells with moisture. Light rain will not, however,

do much harm. Whilst Siebold was endeavouring to remove some of the water from a nest which had been drenched in a shower, he was astonished to find the wasps themselves already busy at the work, putting their heads into the cells, sucking up the water, then passing to the edge of the nest and spitting out the fluid. In this way they are able to get rid of the effects of a wetting if it is not very severe. Though the Polistes feed their young exclusively with animal food, they yet appear to collect a sweet fluid which Siebold found in some cells, and which he thinks the workers take for their own enjoyment, since they were seen entering such cells and apparently sucking at the contents—in fact, taking a little refreshment in the intervals of their labour.

The development of the grub is carefully described by the author, and a "pseudo-nymph" stage is recognised intervening between the nymph and the pupa. The perfect insect bites off the lid of its cell, and comes out with perfect wings, deposits first of all a drop of urinary excretion, and makes a trial flight, then returns to take part in the labours of the colony. The cell is often used again for another egg. The first drones make their appearance with the beginning of July, an important fact for Siebold's experiments, for if the nest is to be used at all now is the latest moment; they have to be killed off, and all the remaining larvæ and pupæ destroyed—in order to secure a colony consisting solely of virgins. The drones play a pitiable part in the nest—sneaking about in the empty cells and behind the comb. Not till the month of August are their generative organs fully developed, and then they make their first approaches to the females. Their proceedings are minutely described, and it appears that they meet with many rebuffs from the busily-occupied workers of the hive, and that it is outside at a distance from the nest that their addresses are at length accepted by those of the larger females destined to become queens. Not all the large females appear to have this destiny, and none appear to leave the nest until all the brood has been brought through, when (about the beginning of October) the nests become deserted. Only a few flattened old virgin wasps remain, who are killed off by the frosts, whilst the young queens have married and sought out for themselves winter quarters. Siebold distinguished black-eyed and green-eyed drones, and speculates upon the signification of this difference.

Having ascertained these and other matters relating to the Polistes in far greater detail than we have been able here to indicate them, Siebold was prepared to make his experiment. In the nest from which he wanted an answer to these questions, " Can unfertilised Polistes females lay eggs which will develop ? " and if so, " Of what sex will the parthenogenetically-produced progeny be ? " he proceeded to destroy the queen and all the eggs, larvæ, and pupæ in the cells with the greatest care as late as possible in the season, so as to have as large a colony as possible left, the limit of the time being given by the date of the appearance of the first drone. The queens thus taken were used for careful histological study of the generative organs, and since in all cases Siebold found the *receptaculum seminis* filled with moving spermatozoa, he was able to feel assured that he had really removed the queen in each case. We will merely direct the attention of those interested in histology to the minute description here given of the ovary, which in the main agrees with Leydig's, and of the *receptaculum seminis*, which in opposition to Leydig, on account of its nerve supply, Siebold holds to be contractile. After waiting some days Siebold had the gratification of finding the first eggs laid in the cells of several of the nests from which he had removed queen, eggs, and larvæ, and he felt convinced that they could only have been laid by the small virgin workers who alone tenanted the combs. The whole business of the colonies proceeded just as well as when the queens were there, and the virgins watched and worked with the same assiduity as had done their queen-mother. In some cases Siebold actually saw a worker deposit an egg, and such egg-laying workers, when anatomically tested, showed, firstly, in the presence of *corpora lutea* (the precise signification of which the investigator had ascertained by his histological studies of the ovary) that eggs had been extruded, and, secondly, in the complete absence of spermatozoa from the *receptaculum seminis*, that the insect was a virgin. Out of a hundred nests which he had begun to observe in one season, and out of one hundred and fifty in another, only some twenty or so in each case came through all the long series of accidents from weather, insects, birds, etc., to which they were necessarily exposed,

and some of those which promised the best results and
had cost the most pains came to a bad end in the
very last days of the research. In order to determine
the sex of the wasps born from the eggs laid by the
parthenogenetic females, Siebold in most cases only
allowed the development to proceed sufficiently far to
enable him to recognise the sex by anatomical investigation. The dried skin, however, of such grubs as
were found dead in their cells afforded sufficient evidence of the sex. In all cases the parthenogenetic
offspring was without exception male. The queen-
wasps, as we have mentioned, also late in the season
lay eggs which produce drones, which are easily distinguished from the drones parthenogenetically produced by their larger size. It occurred to Siebold
when he first ascertained that the queens produce
drones that such drones might visit his virgin colonies,
and thus his whole experiment be invalidated. He
was, however, reassured on this point by a nearer
acquaintance with the Polistes; for such drones are
not born till later than the period at which his small
females laid their eggs, the former event taking place
at the end of July, the latter at the beginning; and,
furthermore, as we have noticed above, it is not till
still later (August), when the experimental cells were
long since all occupied with eggs, that the power and
desire of sexual activity comes to these drones.

Siebold's experiments extended over four years,
and although some hundreds of nests were more or
less observed, only thirty-seven—but these amply
sufficient—gave the answer to his questions, passing

successfully through all the stages above noted.
Firstly, they furnished a virgin colony in a nest absolutely free from eggs and larvæ—except a few
advanced larvæ purposely left in some nests and noted
down—which colony laid eggs; secondly, these eggs
produced without exception (some few eggs not
developing) males.

The method of recording which was used must be
mentioned to give a notion of the accuracy of the
observations. A series of plans of each nest was kept,
each cell being represented and its contents at different dates. Successive plans were used for recording
the successive changes in the number of cells of the
nest, and in their contents at different periods of the
observations. Signs jotted down in the plan cells
indicate such facts as these—*e.g.* the cell contains a
" parthenogenetic egg," or " a second parthenogenetic
egg which was laid after a first one had disappeared,"
or " a larva sprung from the queen," or " a parthenogenetic male larva," etc. etc. A second record was
kept, and is given for twenty-two cases, in which the
following facts were noted: Number of the nest,
date it was made movable, number of cells at that
time, day of emergence of first worker-female, date of
destruction of queen, eggs, and grubs, number of larvæ
and pupæ left undestroyed at this date, date of first
laying of parthenogenetic eggs, date of first emergence of parthenogenetic larvæ, date of first emergence
of drones born from queens' eggs (these were null in
most cases, and were always so late as not to affect
the experiments by possibly impregnating the worker-

females), number of the same, number of cells observed
when the experimental conditions were established,
date and duration of the experiment, maximum number
of female workers employed in the affairs of the nest,
number of larvæ, pupæ, and wasps of the parthenogenetic brood found at the conclusion of the experiment. After the account of the artificially obtained
results, two cases are recorded in which Siebold found
a parthenogenetic colony naturally established by an
accident which had destroyed their queen and comb.

Before concluding this chapter of his book, Siebold
makes the very important observation that the facts
observed in the parthenogenesis of Polistes are in
opposition to the view maintained by Leydig, viz.
that the sexual differentiation of the egg is independent
of its fertilisation, and that the evolution of the male
sex is due to diminution of nutrition and warmth.
Bessels has already, in opposition to Landois, shown
that this is not the case in the bee. If it were true
for Polistes, the eggs laid in the early year, when it
is cold, and when there is only the queen to attend
to the larvæ, should produce drones. On the contrary,
they produce females, and the drones appear precisely
at the time when warmth and nourishment are most
abundant.

Siebold concludes, therefore, that (1) the eggs
bring with them from the ovary the capacity of
differentiating themselves as males, and (2) of developing themselves, independently of male influence
into male individuals; (3) but the same eggs can be
changed in these properties by the influence of the

male sperm elements, and proceed to develop as female individuals.

The second chapter, very short, is on Parthenogenesis in *Vespa holsatica*, which was inferred to occur from the observation of a naturally-produced queenless colony, the larvæ in the cells of which were all male.

The third chapter is on Parthenogenesis in *Nematus ventricosus*, the larva of which is known as the Gooseberry-caterpillar. Since three or more generations of these leaf-wasps occur in the season, they furnished abundant material, and the old supposition of parthenogenesis first put out as regards them by Robert Thorn, in the *Gardener's Magazine*, 1820, is shown by Siebold to be justified by carefully conditioned experiment. Some valuable observations on the anatomy of the generative organs, and on the curious increase in the size of the egg after it is laid, are given. The parthenogenetically produced progeny are in this case also male. The results of the *Nematus* experiments were not ready for publication until after the issue of the present work, and we have received, through the kindness of Dr. Dohrn, a copy of the Sitzungsberichte of the Munich Academy of 4th November 1871, in which they are fully given. It appears that though an occasional female appeared among the male broods produced by unfertilised females, this was, in every case where it happened, fully accounted for by the accidental access of a fertilised female, or some such misadventure, duly noted in the records kept of the observations.

Of the fourth and fifth chapters, treating of Parthenogenesis in the Lepidoptera, *Psyche helix*, *Solenobia triquetrella* and *S. lichenella*, we have not space to speak in detail. The same intimate inquiry, and the same very necessary prodigality in the amount of material subjected to experiment, which we noted above as to Polistes, characterise Professor Siebold's treatment of these cases. The parthenogenesis in these cases produces female broods, and though the male of *Psyche helix* has been discovered since Siebold's former researches on this moth, his conclusion is by no means invalidated, for the males are excessively rare. They were first discovered by Claus, of Marburg, who has indicated characters by which future observers may distinguish the sex of the caterpillars. Out of many hundreds of broods reared by Siebold, taken in various places, ranging from the Baltic to the plains of Lombardy, only once did he obtain males. There appear to be thus broods which are entirely female, and broods which are of mixed sexes. The conditions under which the male sex makes its appearance are not yet ascertained. It is exceedingly desirable that those who may be fortunate enough to come across a mixed brood, should make experiments to ascertain if all the eggs which are fertilised produce males. The females of the purely female broods are completely developed in every respect, having perfect copulatory organs, and the egg is furnished with a micropyle ; therefore, as Siebold maintains, they must not be called pseudova. It should be mentioned that the inquiries necessary to establish the identity

of the species, and the distinctive characters with
regard to these little moths, have occupied a great
deal of our author's time and attention, and are here
recorded.   In regard both to Psyche and Solenobia
examination with the microscope was employed to
determine the absence of male elements from the
receptaculum seminis; and we have moreover an
account of the structure of the ovaries.   In relation
to this matter, Professor Siebold takes the opportunity
of replying to some criticisms of his former work by
M. Plateau, who appears to have made little of the
arguments based on the proof thus obtained of virgin-
ity, without knowing the real extent and nature of
Siebold's studies, having, in fact, only read of them
in an imperfect abstract.   It appears also that M.
Plateau took "ein einziger Fundort" to mean "un
naturaliste collecteur," an amusing mistake to which
our attention is drawn in a note, p. 155.   We may
briefly mention here with regard to Solenobia, that it
appears that *S. lichenella* is only the female brood
of *S. pineti,* of which males and females regularly
occur.   No structural difference appears to exist
between the two kinds of females, but the former,
on escaping from the chrysalis-sac, at once proceed
to lay eggs, which produce invariably females; whilst
the latter wait for copulation, and if that be with-
held, die, and dry up without laying their eggs.
These insects offer most promising material for further
researches on the conditions attending the differentiation of sex.

   We now come to the sixth and last chapter, on

"the Parthenogenetic Reproduction in Apus and
allied Crustacea." Already, in 1856, Siebold had
stated his supposition that *Apus cancriformis*, *Limnadia gigas*, and *Polyphemus oculus*, in which species
no males had been observed, presented examples of
true parthenogenesis, and were not to be regarded as
bud-producing "nurses," in a so-called alternation of
generations. Leuckart subsequently expressed the
same opinion with regard to the reproduction, independent of males, observed in *Daphnia*, *Apus*, and
Limnadia. Ever since that period Siebold has continually kept an eye upon Apus. In 1858 the males
of Apus were discovered by Kozubowski, and Siebold
received specimens from various localities. He thus
learned to distinguish with perfect facility the two
sexes, and was enabled now to convince himself that,
as with the Lepidoptera above spoken of, so with
Apus, broods occur which are entirely destitute of
males, and go on reproducing parthenogenetically,
whilst other broods occur in which both sexes are
present. The number of Apus of two species—*Apus
cancriformis* and *Apus productus*—examined by
Siebold, amounts actually to some thousands. He
received quantities taken from various ponds in middle
Europe (Apus occurs in shallow pools which dry up
during parts of the year, and it can be taken in
immense quantity), and had the opportunity of studying one pond—that at Gossberg, near Munich, with
minuteness, from the year 1864 to the year 1869
inclusive, besides casual examinations of the same
pond in 1857 and 1858. Time after time, taking

several hundreds of the Apus from the pond, he never found a single male amongst them. On one occasion he had the whole contents of the little pond removed with the greatest care, so as to feel sure that he had obtained every Apus present. He received on this occasion 5796 specimens of Apus, *every one of which being carefully examined* proved to be female. At the same time 2576 specimens of Branchipus were obtained from the pond, which were, as usual, of both sexes. In those cases where ponds afforded both males and females of Apus, it is remarkable that the proportion of the sexes was very variable. The highest proportion of males appears to be in a case recorded by Sir John Lubbock, who found thirty-three male and thirty-nine female *Apus productus* in a pond near Rouen, whilst among 193 specimens of *Apus cancriformis*, from a locality near Krakow, only one male occurred. What is most important about this variation in the proportion of males to females is that in two or three localities, furnishing mixed generations of Apus, from which he has received, year after year, numbers of specimens, Siebold has observed an apparent constantly-augmenting dispro-portion of males to females, and he is led to the supposition that in these cases the males will at last cease altogether, and thus a female generation be pro-duced which will continue to reproduce itself partheno-genetically, as in the Gossberg and a great number of other ponds. This is, however, by no means proved; and we have no idea at present as to how the males may make their appearance again, or what are the conditions affecting their development and extinction. It occurred to Siebold that an objection might be urged against parthenogenesis in Apus, in that, although he examined consecutive generations and found them always female, he could not be sure that males had not been present before he took his specimens, and had not died and decomposed after having fertilised the females. To meet such an objection, he firstly made himself thoroughly acquainted with the male generative organs and the spermatozoa, and secondly with the ovaries and their development. He found the spermatozoa to be motionless like those of other Crustacea, and he never succeeded in detecting any of them in the female genitalia amongst the specimens belonging to supposed female generations. But he equally failed to find spermatozoa or a receptacle for them in the female genitalia of the specimens of mixed generations, and therefore no conclusion could be drawn from the observation. The structure and development of the ovum, however, made this observation decisive, since it was found that an eggshell forms round the ovum in the uterus, and, in the absence of a micropyle, fertilisation, if it takes place at all, must be accomplished before this shell is hardened. A further proof of another kind was obtained by experiment. Having removed eggs from females, which certainly at the time contained no spermatozoa, Siebold placed them in a small tank, and from these obtained Apus-embryos. Others were reared to maturity from eggs taken in the pond.

The relative size of male and female is a question

about which there is some interest; differences which
have been observed seem to depend on this, that
Apus continues growing as long as the pond in which
it lives does not dry up, and hence the eggs which
hatch soonest give the largest-sized progeny.   In his
tabular statements Siebold gives measurements of the
specimens examined by him at different times from
various localities.

A few words must be said here upon the very
extraordinary history of the ovum of Apus made out
by Siebold, the structures being identical, whether
the female examined belonged to a parthenogenetic
or digenetic brood.   The essential female organs of
reproduction in Apus may be roughly described as
two large tubes placed on either side of the alimentary
canal, opening externally at the posterior end, and
giving off towards the other end primary and secondary branches.   On the ends of these short secondary
branches are situated the egg follicles.   Four cells
appear in each egg follicle in a very early stage of its
development, and one of these takes on more rapid
growth—becoming the egg-cell—whilst the others
disappear as deutoplasmogen or vitellogenous cells;
the egg then acquires some size and a red colour, and
has a visible germinal vesicle.   But such eggs are
much smaller than the eggs observable in the main
stem of the ovarian tube, and this appears to be the
very startling explanation.   The eggs escape from
their follicles as a matter of course, and pass along
the canal leading from it to a primary branch of the
ovarian tube, and there two and sometimes three of

these eggs *fuse into one mass,* around which a shell is secreted, and which thus forms the actual egg—really a threefold egg ; and from such a wonderfully formed egg only one embryo develops. Unfortunately we are not told what becomes of the germinal vesicles ; according to the drawings they seem to disappear at this stage. Siebold appears to have convinced himself that the fusion is a normal thing, and not due to any pressure or osmotic action taking place during the microscopical examination. The structure of the ovary of Apus is figured in a plate.

As to the other crustaceans named, which are *Artemia salina* and *Limnadia Hermanni,* the occurrence of parthenogenetic broods is inferred from the descriptions of other writers whose works are criticised at some length, and also from examination of specimens. It seems not impossible from an observation of Zenker that in *Artemia salina* parthenogenetic alternate with digenetic broods. In the beginning of the year 1851 this observer found three males among one hundred females, later in July the same pond furnished thousands of females, but not one male.

In conclusion, Professor Siebold, whilst adopting Leuckart's term "Arrenotoky," to designate the phenomenon of the parthenogenetic production of male offspring, as seen in the Hymenoptera, proposes the parallel term, "Thelytoky," for the parthenogenetic production of female offspring as demonstrated now conclusively in some Lepidoptera and Crustacea. It seems to us that a third term should also be available for the case of mixed offspring (that is, of two

sexes) such as "Amphotoky"; and the terms need
not be limited to parthenogenetic cases. In his concluding remarks, whilst repeating the expression of
his conviction that parthenogenesis will be found
more and more to be of frequent and fixed occurrence
in various classes of animals, Siebold alludes with
caution to the list of cases in which parthenogenesis
is stated to occur, given by Gerstaecker in Bronn's
Classen und Ordnungen des Thierreichs. Gerstaecker rightly enough distinguishes cases in which
parthenogenesis has been observed as an accidental
and rare exception, and those in which it has a
definitely recurring place. Siebold considers (and
after the great pains he has himself expended on the
cases recorded in this book, he is fully warranted in
so doing) that many of the examples put forward by
Gerstaecker require a more careful testing, and he
offers some remarks on parthenogenesis in the gall-
flies, and in the silkworm moth. Finally, he alludes
to cases among Vertebrates in which indications of a
power of development in the egg, independent of the
male element, have been observed. The most remarkable of these is that quoted by Leuckart in his work
already cited, which Siebold omits here, but has done
justice to in the short supplementary paper read at
the Munich Academy since the publication of this book.
In 1844 Professor Bischoff found ova in the uterus
of an unimpregnated sow, which exhibited segmenta-
tion of the yelk, some into two and four, and others
into sixteen and twenty divisions. Other cases here
given are as follows : In the oviduct of a three-year-old

rabbit, thoroughly separated pathologically from the
uterus, Professor v. Hensen of Kiel found ova in various
stages of yelk-division, and some of their cells had
even advanced into a branched condition. Dr.
Oellacher of Innsbruck has observed stages of yelk-
division in unfertilised hen's eggs. In fishes, in
1859, Agassiz observed yelk-division occurring in the
eggs of Gadidæ, whilst yet in the ovary, and considered it to be due to impregnation, even stating
that he had seen certain fishes place themselves in
such a position as to favour this supposed intra-
ovarian fertilisation. Burnett has since investigated
the case, and concludes that the yelk-division is
independent of fertilisation, a supposition which is
rendered in every way probable from other researches
on the fish egg; but, curiously enough, Dr. Burnett
thinks these eggs should be regarded as "germs," and
not as "true eggs," an opinion to which Siebold, of
course, is completely opposed, and one which, in in-
vertebrate cases, has been shown to be untenable.

Siebold does not allude to those cases of ovarian
cysts found occasionally in the unfertilised human
female, and containing hair and teeth—a phenomenon
which we should be glad to see further discussed and
investigated, since, as far as we can remember, the
origin of the contents of such cysts from irregularly
developing ova is probable. The eel is suggested as
a possible parthenogenetic vertebrate. It is a very
strange fact that we are still ignorant of the ripe eggs
and embryos as well as of the males of the eel, even
as in the time of Aristotle. With the following words

of that great naturalist, addressing them to those
who still refuse to accept the existence of Parthenogenesis, Siebold ends his book : " More belief must
be given to observation than to theory, and this last
is only worthy of belief when leading to the same
result as experience."

VIII

A THEORY OF HEREDITY

A Review of Haeckel's Pamphlet entitled *Perigenesis der Plastidule*. From *Nature*, 15th July 1876.

UNDER the title "Perigenesis der Plastidule oder die Wellenzeugung der Lebenstheilchen," Professor Haeckel has published quite recently a pamphlet containing an attempt to furnish a mechanical explanation of the elementary phenomena of reproduction which shall be more satisfactory than Mr. Darwin's ingenious and well-known theory of Pangenesis. I shall endeavour to show that Professor Haeckel's theory is essentially that with which both English and German students of Mr. Herbert Spencer's works have long been familiar; and that it does not furnish a clearer explanation than does Mr. Darwin's Pangenesis, of the special facts of heredity which Mr. Darwin had in view.

Haeckel commences with a very concise statement of what is at present known as to the visible composition of " plastids," those corpuscles of life-stuff called " cells " by Schleiden and Schwann, "elementary organisms " by Brücke, " life-units " by Darwin. Most plastids possess a differentiated central kernel or nucleus, which again possesses one or more nucleoli. The substance of which the body of such a nucleated plastid or *true* cell is mainly composed is generally known by von Mohl's term, " protoplasma." Haeckel proposes to distinguish the substance of the nucleus

T

by the name "coccoplasma." In the simplest form of plastid, the "cytod," which is devoid of nucleus, and is exhibited by those lowly organisms known as Monera, by the young Gregarina (Ed. van Beneden), by the hyphæ of some Fungi, and by the ripe egg of all organisms (if we may judge from the results of the most recent researches), coccoplasm and protoplasm are not differentiated, but exist as one substance, which Haeckel, following Ed. van Beneden, distinguishes as "plasson." Whether these distinctions have a real value or not, is of no moment for the question in hand. It is a widely-accepted doctrine—in fact, the fundamental generalisation on which Biology as a science rests—that the excessively complex chemical compound which forms the substance of plastids or life-units is the ultimate seat of those phenomena or manifestations of energy which distinguish living from lifeless things—to wit, growth by intus-susception, reproduction, adaptation, and continuity or hereditary transmission. Leaving Professor Haeckel's pamphlet for a time, let us go back thirteen years.

As long ago as July 1863, Mr. Herbert Spencer, in his *Principles of Biology*, pointed out at considerable length (vol. i. p. 181) that the assumption of definite forms, and the power of repair exhibited by organisms, is only to be brought into relation with other facts (that is to say, so far explained) by the assumption that *certain units* composing the living substance or protoplasm of cells possess "polarity" similar to, but not identical with, that of the units which build up crystals. Mr. Spencer is careful to explain that by the term "polarity" we mean simply to avoid a circuitous expression, namely, the still unexplained power which these units have of arranging themselves into a special form. He then points out that the units in question cannot be the molecules of the proximate chemical compounds which we obtain from protoplasm—such as albumen, or fibrin, or gelatin, or even protein. Further he shows that they cannot be the *cells* or morphological units, since such organisms as the Rhizopods are not built up of cells, and since, moreover, "the formation of a cell is to some extent a manifestation of the peculiar power" under consideration. "If then," he continues, "this organic polarity can be possessed neither by the chemical units, nor the morphological units, we must conceive it as possessed by certain intermediate units, which we may term *physiological*. There seems no alternative but to suppose that the chemical units combine into units immensely more complex than themselves, complex as they are; and that in each organism, the physiological units produced by this further compounding of highly compound atoms, have a more or less distinctive character. We must conclude that in each case, some slight difference of composition in these units, leading to some slight difference in their mutual play of forces, produces a difference in the form which the aggregate of them assumes."

Further on Mr. Spencer applies the hypothesis of physiological units to the explanation of the phenomena of heredity, introducing the subject by the following admirable remarks, which appear to me to assign in the most judicious manner their true value to such hypotheses and to be as strictly applicable to later speculations as to his own. " A positive explanation of heredity is not to be expected in the present state of biology. We can look for nothing beyond a simplification of the problem, and a reduction of it to the same category with certain other problems which also admit of hypothetical solution only. If an hypothesis which certain other widespread phenomena have already thrust upon us, can be shown to render the phenomena of heredity more intelligible than they at present seem, we shall have reason to entertain it. The applicability of any method of interpretation to two different but allied classes of facts is evidence of its truth. The power which organisms display of reproducing lost parts, we saw to be inexplicable except on the assumption that the units of which any organism is built have an innate tendency to arrange themselves into the shape of that organism. We inferred that these units must be the possessors of special polarities, resulting from their special structures ; and that by the mutual play of their polarities they are compelled to take the form of the species to which they belong. And the instance of the *Begonia phyllomaniaca* left us no escape from the admission that the ability thus to arrange themselves is latent in the units in every undifferentiated cell. . . . The assumption to which we seem driven by the *ensemble* of the evidence, is that sperm-cells and germ-cells

are essentially nothing more than vehicles, in which
are contained small groups of the physiological units
in a fit state for obeying their proclivity towards the
structural arrangement of the species they belong to.
. . . If the likeness of offspring to parents is thus
determined, it becomes manifest, *a priori*, that besides
the transmission of generic and specific peculiarities,
there will be a transmission of those individual peculiarities which, arising without assignable causes, are
classed as 'spontaneous.' . . .

"That changes of structure caused by changes of
action must also be transmitted, however obscurely,
from one generation to another, appears to be a deduction from first principles—or if not a specific
deduction, still, a general implication. . . . Bringing
the question to its ultimate and simplest form, we may
say that as on the one hand physiological units will,
because of their special polarities, build themselves
into an organism of a special structure, so on the other
hand, if the structure of this organism is modified by
modified function, it will impress some corresponding
modification on the structures and polarities of its
units.   The units and the aggregate must act and
react on each other.   The forces exercised by each
unit on the aggregate, and by the aggregate on each
unit, must ever tend towards a balance.   If nothing
prevents, the units will mould the aggregate into a
form in equilibrium with their pre-existing polarities.
If contrariwise, the aggregate is made by incident
actions to take a new form, its forces must tend to
re-mould the units into harmony with this new form ;

and to say that the physiological units are in any degree so re-moulded as to bring their polar forces towards equilibrium with the forces of the modified aggregate, is to say that when separated in the shape of reproductive centres, these units will tend to build themselves up into an aggregate modified in the same direction." (P. 256.)

Thus, then, Mr. Herbert Spencer definitely assumes an order of molecules or units of protoplasm—lower in degree than the visible cell-units or plastids—to the "polar forces" of which and their modification by external agencies and interaction, he ascribes the ultimate responsibility in reproduction, heredity, and adaptation.

I am unable to say whether Mr. Darwin was acquainted with or had considered Mr. Herbert Spencer's hypothesis of physiological units, when in 1868 he published his own provisional hypothesis of Pangenesis. But an examination of the bearings of the two hypotheses shows that the former does not render the latter superfluous, nor is the one inconsistent with the other. Mr. Darwin wished to picture to himself and to enable others to picture to themselves a process which would account for (that is, hold together and explain) not merely the simpler facts of hereditary transmission, but those very curious though abundant cases in which a character is transmitted in a latent form and at last reappears after many generations, such cases being known as "atavism" or "reversion"; and again those cases of latent transmission in which characteristics special to the male are transmitted to the male off-

spring through the female parent without being manifest in her; and yet again the appearance at a particular period of life of characters inherited and remaining
latent in the young organism.   According to the
hypothesis of pangenesis, "every unit or cell of the
body throws off gemmules or undeveloped atoms,
which are transmitted to the offspring of both sexes
and are multiplied by self-division.   They may remain
undeveloped during the early years of life or during
successive generations; their development into units
or cells, like those from which they were derived,
depending on their affinity for, and union with, other
units or cells previously developed in the due order of
growth."

In an essay ("Comparative Longevity," Macmillan,
1870, p. 32) published six years ago, I briefly suggested the possibility of combining Mr. Herbert
Spencer's and Mr. Darwin's hypotheses thus: "The
persistence of the same material gemmule and the
vast increase in the number of gemmules, and consequently of material bulk,[1] make a *material* theory
difficult.   Modified force-centres, becoming further
modified in each generation, such as Mr. Spencer's
physiological units, might be made to fit in with Mr.
Darwin's hypothesis in other respects."   In fact, in
place of the theory of emission from the constituent
cells of an organism of material gemmules which circulate through the system and affect every living cell,

---

[1] On this subject see Mr. Sorby's recent Presidential Address to the
Royal Microscopical Society, in the *Quarterly Journal of Microscopical
Science*, April 1876.

and accumulate in sperm-cells and germ-cells, we may substitute the theory of transmission of force, the two theories standing to one another in the same relation as the emission and undulatory theories of light.

It may, however, be very fairly questioned whether our conceptions of the vibrations of complex molecules, or in other words, their force-affections, are sufficiently advanced to render it desirable to substitute the vaguer though possibly truer undulatory theory of heredity for the more manageable molecular theory (Pangenesis). How are we to conceive of the propagation of such states of force-affection or vibration (as they are vaguely termed) through the organism from unit to unit? In what manner, again, are we to express the dormancy of the pangenetic gemmules in terms of molecular vibration? It is true that molecular physics furnishes us with some analogies in the matter of the propagation of particular states of force-affection from molecule to molecule, as, for example, in the various modes of decomposition exhibited by gun-cotton, in contact actions and the like; but it will require a very extended analysis of both the phenomena of heredity and of molecular phenomena similar to those just cited, to enable us to supersede the admittedly provisional hypothesis of Pangenesis by a hypothesis of vibrations. And it is necessary here to remark that in the fundamental conception of Pangenesis, namely, the detachment from the living cells of the organism of gemmules which then circulate in the organism, there is nothing contrary to analogy, but rather in accordance with it.

It is quite certain that in some infective diseases the contagion is spread by specific material particles. This seems to be established, although it is far from settled as to whether these particles are parasitic organisms or portions of the diseased organism itself. Mr. Darwin's pangenetic gemmules may, even if not accumulated and transmitted from generation to generation, be called upon to explain the solidarity of the constituent cells of one organism ; they may be assumed as agents of a peculiar kind of infection,[1] by means of which the molecular condition or force-affection of one cell is communicated to others at a distance in the same organism. It is difficult without some such hypothesis of an active material exchange of living molecules between the various cells of the body, to conceive of the way in which "change is propagated throughout the parental system," or a modified part is to "impress some corresponding modification on the structures and polarities" of distant units, such, for example, as those contained in the mammalian ovum.

In the human ovary no egg-cells are produced after the age of two and a half years. Each of the many hundred eggs there contained reposes quietly in its follicle, whilst the growth and development of other organs is proceeding. Then a renewed period of activity for the ovary commences, but the majority

[1] It is a striking exemplification of the unity of biological science that we should have to look to the pathologist for the next step in this region of speculation, and that fermentations, phosphorescence, fevers, and heredity, should be simultaneously studied from a common point of view with psychology.

of the originally-formed egg-cells retain their vitality
and form-individuality for more than forty years.
How, we may ask, during that time are they subjected
to the influence of new polar forces acquired by the
other units of the body? We know that they are so
impressed, or have such influences propagated to them.[1]
Is it by "action at a distance," or by the contact action
of circulating infective gemmules?

Such being the state of speculation, in England at
any rate, with regard to the mechanical explanation
of heredity, we return to Professor Haeckel's recently
enunciated theory of the Perigenesis of plastidules.

It is clear, to begin with, that Professor Haeckel
has either never studied or has forgotten Mr. Herbert
Spencer's writings. His attempt to substitute something better for Mr. Darwin's provisional hypothesis
of Pangenesis, as he tells us, has its origin, to a great
extent, in the admirable popular lecture of Professor
Ewald Hering of Prague, " Über das Gedächtniss als
eine allgemeine Function der organisirten Materie "
[On Memory as a General Function of Organised
Matter], published in 1870, and to some extent,
including terminology, is based on an essay by Elsberg,
of New York, published in the *Proceedings* of the
American Association, Hartford, 1874. With the latter
of these publications I am only acquainted through
Professor Haeckel's citations, but with the former at
first hand. Professor Hering gives a brief outline, in
the lecture in question, of the fundamental doctrine
of physiological psychology, which had been previously

---

[1] See note on page 286.

worked out to its consequences on an extensive scale by
Mr. Herbert Spencer. Professor Hering has the merit
of introducing some striking phraseology into his
treatment of the subject, which serves to emphasise
the leading idea. He points out that since all transmission of "qualities" from cell to cell in the growth
and repair of one and the same organ, or from parent
to offspring, is a transmission of vibrations or affections
of material particles, whether these qualities manifest
themselves as form, or as a facility for entering upon
a given series of vibrations, we may speak of all such
phenomena as "memory," whether it be the conscious
memory exhibited by the nerve-cells of the brain or
the unconscious memory we call habit, or the inherited
memory we call instinct; or whether again it be the
reproduction of parental form and minute structure.
All equally may be called "the memory of living
matter." From the earliest existence of protoplasm
to the present day, the memory of living matter is
continuous. Though individuals die, the universal
memory of living matter is still carried on.

Professor Hering, in short, helps us to a comprehensive conception of the nature of heredity and adaptation
by giving us the term "memory," conscious or unconscious, for the continuity of Mr. Herbert Spencer's
polar forces or polarities of physiological units.

Elsberg appears (though this is only an inference
on my part) to be acquainted with Mr. Herbert
Spencer's hypothesis of physiological units. Adopting Haeckel's useful term "plastid" for a corpuscle of
protoplasm (cell or cytod), he designates the physio-

logical units "plastidules," a name which Haeckel has
accepted, and which may very possibly be found permanently useful. But Elsberg does not appear to
have helped on the discussion of the subject to a great
extent, since he proceeds no further than is implied
in adopting Mr. Darwin's theory of Pangenesis, whilst
substituting the "plastidules" for Mr. Darwin's
"gemmules." It appears to me that Elsberg in his
combination of the Spencerian and Darwinian hypotheses, has omitted the sound element in the latter.
and retained the more questionable. He should have
conjoined Mr. Herbert Spencer's conception of "plastidules"—possessing special polarities or force affections
which they are capable of propagating as *changes* of
state (*i.e.* force-waves) to associated plastidules, and
so to offspring—with Mr. Darwin's conception of a
universal and continuous emission of such changes
from all the cells of an organism, and the frequent
occurrence of a persistently latent condition of those
changes—a condition which Hering's happy use of
the term "memory" enables us to illustrate by the
analogous (or we should rather say identical) "latent"
or "dormant condition" of mental impressions.

This is, in fact, the position which Professor Haeckel
takes up—though independently of what Mr. Spencer
has written on the subject, excepting so far as the
influence of the latter is to be traced in Elsberg's
essay. For Haeckel, living matter, protoplasm, or
plasson consists of definite molecules—the plastidules
—which cannot be divided into smaller plastidules,
but can only be split into lower chemical compounds.

What Mr. Spencer calls polarities or polar forces
Haeckel speaks of as "undulatory movements"—a
symbol which has the advantages and disadvantages
of analogy, but which, like "polarity," is only a
symbol, and covers our incapability of conceiving
more definitely the character of the phenomenon it
designates. The undulatory movement of the plastidules is the key to the mechanical explanation of all
the essential phenomena of life. The plastidules are
liable to have their undulations affected by every
external force, and once modified the movement does
not return to its pristine condition. By assimilation
they continually increase to a certain point in size,
and then divide, and thus perpetuate in the undulatory
movement of successive generations the impressions
or resultants due to the action of external agencies on
individual plastidules. This is Memory. All plastidules possess memory—and Memory, which we see in
its ultimate analysis is identical with reproduction, is
the distinguishing feature of the plastidule ; is that
which it alone of all molecules possesses in addition
to the ordinary properties of the physicist's molecule ;
is in fact that which distinguishes it as vital. To the
sensitiveness of the movement of plastidules is due
Variability—to their unconscious Memory the power
of Hereditary Transmission. As we know them to-day,
they may "have learnt little and forgotten nothing"
in one organism, "have learnt much and forgotten
much" in another, but in all, their Memory, if sometimes fragmentary, yet reaches back to the dawn of
life on the earth.

NOTE

THE two preceding essays give an account of researches and speculations which have a historical value for the philosophic biologist. In both of these articles the reader will find evidence of a belief on my part in "the transmission by heredity of acquired characters." This belief is one which I now think is not justifiable. Some of the questions connected both with Parthenogenesis and Heredity have been elucidated in a masterly manner by August Weismann of Freiburg, whose essays recently published by the Oxford University Press cannot be too carefully studied by naturalists.

15TH MARCH 1890.

IX

THE HISTORY AND SCOPE
OF ZOOLOGY

From the Ninth Edition of the *Encyclopædia Britannica*

THE branch of science to which the name Zoology is strictly applicable may be defined as that portion of Biology which relates to animals, as distinguished from that portion which is concerned with plants.

The science of Biology itself has been placed by Mr. Herbert Spencer in the group of concrete sciences, the other groups recognised by that writer being the "abstract-concrete" and the "abstract." The abstract sciences are Logic and Mathematics, and treat of the blank forms in which phenomena occur in relation to time, space, and number. The abstract-concrete sciences are Mechanics, Physics, and Chemistry. The title assigned to them is justified by the fact that, whilst their subject-matter is found in a consideration of varied concrete phenomena, they do not aim at the explanation of complex concrete phenomena as such, but at the determination of certain "abstract" quantitative relations and sequences known as the "laws" of mechanics, physics, and chemistry, which never are manifested in a pure form, but always are inferred by observation and experiment upon complex phenomena in which the abstract laws are disguised by their simultaneous interaction. The group of concrete sciences includes Astronomy, Geology, Biology, and Sociology. These sciences have for their aim to "explain" the concrete complex phenomena of (a) the sidereal system, (b) the earth as a whole, (c) the living matter on the earth's surface, (d) human society, by reference to the properties of matter set forth in the generalisations or laws of the abstract-concrete sciences, i.e. of mechanics, physics, and chemistry.

The classification thus sketched exhibits, whatever its practical demerits, the most important fact with regard to Biology, namely, that it is the aim or business of those occupied with that branch of science to assign living things, in all their variety of form and activity, to the one set of forces recognised by the physicist and chemist. Just as the astronomer accounts for the heavenly bodies and their movements by the laws of motion and the property of attraction, as the geologist explains the present state of the earth's crust by the long-continued action of the same forces which at this moment are studied and treated in the form of "laws" by physicists and chemists, so the biologist seeks to explain in all its details the long process of the evolution of the innumerable forms of life now existing, or which have existed in the past, as a necessary outcome, an automatic product, of these same forces.

Science may be defined as the knowledge of causes; and, so long as Biology was not a conscious attempt to ascertain the causes of living things, it could not be rightly grouped with other branches of science. For a very long period the two parallel divisions of

Biology,—Botany and Zoology,—were actually limited to the accumulation of observations, which were noted, tabulated, and contemplated by the students of these subjects with wonder and delight, but only to a limited extent and in restricted classes of facts with any hope or intention of connecting the phenomena observed with the great nexus of physical sequence or causation. A vague desire to assign the forms and the activities of living things in all their variety to general causes has always been present to thoughtful men from the earliest times of which we have record, but the earlier attempts in this direction were fantastic in the extreme; and it is the mere truth that, at the time when the phenomena of inorganic nature had been recognised as the outcome of uniform and constant properties capable of analysis and measurement, living things were still left hopelessly out of the domain of explanation, the earlier theories having been rejected and nothing as yet suggested in their place.

The history of Zoology as a science is therefore the history of the great biological doctrine of the evolution of living things by the natural selection of varieties in the struggle for existence,—since that doctrine is the one medium whereby all the phenomena of life, whether of form or function, are rendered capable of explanation by the laws of physics and chemistry, and so made the subject-matter of a true science or study of causes. A history of Zoology must take account of the growth of those various kinds of information with regard to animal life which have been arrived at in past ages through the labours of a long

series of ardent lovers of nature, who in each succeeding period have more and more carefully and accurately
tested, proved, arranged, and tabulated their knowledge, until at last the accumulated lore of centuries
—without the conscious act of its latest heirs and
cultivators—took the form of the doctrine of descent
and the filiation of the animal series.

General Historical Sketch

There is something almost pathetic in the childish
wonder and delight with which mankind in its earlier
phases of civilisation gathered up and treasured stories
of strange animals from distant lands or deep seas,
such as are recorded in the Physiologus, in Albertus
Magnus, and even at the present day in the popular
treatises of Japan and China.   That omnivorous universally credulous stage, which may be called the
"legendary," was succeeded by the age of collectors
and travellers, when many of the strange stories believed in were actually demonstrated as true by the
living or preserved trophies brought to Europe by
adventurous navigators.   The possibility of verification established verification as a habit; and the
collecting of things, instead of the accumulating of
reports, developed a new faculty of minute observation.   The early collectors of natural curiosities were
the founders of zoological science, and to this day the
naturalist-traveller and his correlative, the museum
curator and systematist, play a most important part
in the progress of Zoology.   Indeed, the historical and
present importance of this aspect or branch of zoolo-

gical science is so great that the name " Zoology " has until recently been associated entirely with it, to the exclusion of the study of minute anatomical structure and function which have been distinguished as Anatomy and Physiology.   It is a curious result of the steps of the historical progress of the two divisions of biological science that, whilst the word " Botany " has always been understood, and is at the present day understood, as embracing the study, not only of the external forms of plants, their systematic nomenclature and classification, and their geographical distribution, but also the study of their minute structure, their organs of nutrition and reproduction, and the mode of action of the mechanism furnished by those organs, the word " Zoology " has been limited to such a knowledge of animals as the travelling sportsman could acquire in making his collections of skins of beasts and birds, of dried insects and molluscs' shells, and such a knowledge as the museum curator could acquire by the examination and classification of these portable objects.   Anatomy and the study of animal mechanism, animal physics, and animal chemistry, all of which form part of a true Zoology, have been excluded from the usual definition of the word by the mere accident that the zoologist of the last three centuries has had his museum but has not had his garden of living specimens as the botanist has had ;[1] and, whilst the

[1] The mediæval attitude towards both plants and animals had no relation to real knowledge, but was part of a peculiar and in itself highly interesting mysticism.   A fantastic and elaborate doctrine of symbolism existed which comprised all nature ; witchcraft, alchemy, and medicine were its practical expressions.   Animals as well as plants

zoologist has thus for a long time been deprived of the means of anatomical and physiological study—only supplied within the past century by the method of preserving animal bodies in alcohol—the demands of medicine for a knowledge of the structure of the human animal have in the meantime brought into existence a separate and special study of human Anatomy and Physiology.

From these special studies of human structure the knowledge of the anatomy of animals has proceeded, the same investigator who had made himself acquainted with the structure of the human body desiring to compare with the standard given by human anatomy the structures of other animals. Thus Comparative Anatomy came into existence as a branch of inquiry apart from Zoology, and it is only now, in the latter part of the nineteenth century, that the limitation of the word "Zoology" to a knowledge of animals which expressly excludes the consideration of their internal structure has been rejected by the general consent of those concerned in the progress of science; it is now generally recognised that it is mere tautology to speak of Zoology *and* Comparative Anatomy, and that our museum naturalists must give attention as well to the inside as to the outside of animals.

The Anatomy and Physiology of plants have never been excluded from the attention of botanists, because

were regarded as "simples" and used in medicine, and a knowledge of them was valued from this point of view. Plants were collected and cultivated for medicinal use; hence the physic gardens and the botanist's advantage.

in contrast to the earlier zoologists they always were
in possession of the whole living plant, raised from
seed if need be in a hothouse, instead of having only
a dried skin, skeleton, or shell. Consequently the
study of vegetable Anatomy and Physiology has grown
up naturally and in a healthy way in strict relation
to the rest of botanical knowledge, whilst animal Anatomy and Physiology have been external to Zoology
in origin, the product of the medical profession and,
as a consequence, subjected to a misleading anthropo-
centric method.

Whilst we may consider the day as gone by in
which Zoology could be regarded as connoting solely
a special museum knowledge of animals (as twenty-
five years ago was still the case), it is interesting to
observe by the way the curious usurpation of the
word " Physiology," which, from having a wide connotation, indicated by its etymology,—the *physiologus*
of the Middle Ages being nothing more nor less than
the naturalist or student of nature—has in these later
days acquired a limitation which it is difficult to
justify or explain. Physiology to-day means the
study of the physical and chemical properties of the
animal or vegetable body, and is even distinguished
from the study of structure and strictly confined to
the study of function. It would hardly be in place
here to discuss at length the steps by which Physiology became thus limited, any more than to trace
those by which the words "physician" and "physicist"
(which both mean one who occupies himself with
nature) have come to signify respectively a medical

practitioner and a student of the laws of mechanics, heat, light, and electricity (but not of chemistry), whilst the word "naturalist" is very usually limited to a lover and student of living things, to the exclusion of the so-called physicist, the chemist, and the astronomer. It is probable that Physiology acquired its present significance, viz. the study of the properties and functions of the tissues and organs of living things, by a process of external attraction and spoliation which gradually removed from the original *physiologus* all his belongings and assigned them to newly-named and independently-constituted sciences, leaving at last, as a residuum to which the word might still be applied, that medical aspect of life which is concerned with the workings of the living organism regarded as a piece of physico-chemical apparatus.

Whatever may be the history of the word "Physiology," we find Zoology, which really started in the sixteenth century with the awakening of the new spirit of observation and exploration, for a long time running a separate course uninfluenced by the progress of the medical studies of Anatomy and Physiology. The history of every branch of science involves a recognition of the history, not only of other branches of science, but of the progress of human society in every other relation. The century which destroyed the authority of the Church, witnessed the discovery of the New World, and in England produced the writings of Francis Bacon is rightly regarded as the starting-point of the modern knowledge of natural causes or science. The true history of Zoology as a science lies within the three last centuries; and, whilst the theories and fables which were current in earlier times in regard to animal life and the various kinds of animals form an important subject of study from the point of view of the history of the development of the human mind, they really have no bearing upon the history of scientific Zoology. The great awakening of western Europe in the sixteenth century led to an active search for knowledge by means of observation and experiment, which found its natural home in the universities. Owing to the connection of medicine with these seats of learning, it was natural that the study of the structure and functions of the human body and of the animals nearest to man should take root there; the spirit of inquiry which now for the first time became general showed itself in the anatomical schools of the Italian universities of the sixteenth century, and spread fifty years later to Oxford.

In the seventeenth century the lovers of the new philosophy, the investigators of nature by means of observation and experiment, banded themselves into academies or societies for mutual support and intercourse. It is difficult to exaggerate the importance of the influence which has been exercised by these associations upon the progress of all branches of science and of Zoology especially. The essential importance of academies is to be found, as Laplace, the great French astronomer, has said, "in the philosophic spirit which develops itself in them and spreads itself from them as centres over an entire nation and all relations. The isolated man of science can give himself up to dogmatism without restraint; he hears contradictions only from afar. But in a learned society the enunciation of dogmatic views leads rapidly to their destruction, and the desire of each member to convince the others necessarily leads to the agreement to admit nothing excepting what is the result of observation or of mathematical calculation."

The first founded of surviving European academies, the Academia Naturæ Curiosorum (1651),[1] especially confined itself to the description and illustration of the structure of plants and animals; eleven years later (1662) the Royal Society of London was incorporated by royal charter, having existed without a name or fixed organisation for seventeen years previously (from 1645). A little later the Academy of Sciences of Paris was established by Louis XIV. The influence of these great academies of the seventeenth century on the progress of Zoology was precisely to effect that bringing together of the museum-men and the physicians or anatomists which was needed for further development. Whilst the race of collectors and systematisers culminated in the latter part of the eighteenth century in Linnæus, a new type of student made its appearance in such men as John Hunter and other anatomists, who, not satisfied with the superficial observations of the popular "zoologists," set themselves to work to examine anatomically the whole animal kingdom, and to classify its members by aid of the results of such profound study. From them we pass to the compara-

---

[1] The Academia Secretorum Naturæ was founded at Naples in 1560, but was suppressed by the ecclesiastical authorities.

tive anatomists of the nineteenth century and the
introduction of the microscope as a serious instrument
of accurate observation.

The influence of the scientific academies and the
spirit in which they worked in the seventeenth century
cannot be better illustrated than by an examination of
the early records of the Royal Society of London. The
spirit which animated the founders and leaders of that
society is clearly indicated in its motto " Nullius in
verba." Marvellous narrations were not permitted at
the meetings of the society, but solely demonstrative
experiments or the exhibition of actual specimens.
Definite rules were laid down by the society for its
guidance, designed to ensure the collection of solid
facts and the testing of statements embodying novel
or remarkable observations. Under the influence of
the touchstone of strict inquiry set on foot by the
Royal Society, the marvels of witchcraft, sympathetic
powders, and other relics of mediæval superstition
disappeared like a mist before the sun, whilst accurate
observations and demonstrations of a host of new
wonders accumulated, amongst which were numerous
contributions to the anatomy of animals, and none perhaps more noteworthy than the observations, made by
the aid of microscopes constructed by himself, of Leeuwenhoek, the Dutch naturalist (1683), some of whose
instruments were presented by him to the society.

It was not until the nineteenth century that the
microscope, thus early applied by Leeuwenhoek, Malpighi, Hook, and Swammerdam to the study of animal
structure, was perfected as an instrument, and accom-

plished for Zoology its final and most important
service. The earlier half of the nineteenth century is
remarkable for the rise, growth, and full development
of a new current of thought in relation to living
things, expressed in the various doctrines of development which were promulgated, whether in relation to
the origin of individual animals and plants or in relation to their origin from predecessors in past ages.
The perfecting of the microscope led to a full comprehension of the great doctrine of cell-structure and the
establishment of the facts—(1) that all organisms are
either single corpuscles (so-called cells) of living
material (microscopic animalcules, etc.) or are built up
of an immense number of such units; (2) that all
organisms begin their individual existence as a single
unit or corpuscle of living substance, which multiplies
by binary fission, the products growing in size and
multiplying similarly by binary fission; and (3) that
the life of a multicellular organism is the sum of the
activities of the corpuscular units of which it consists,
and that the processes of life must be studied in and
their explanation obtained from an understanding of
the chemical and physical changes which go on in each
individual corpuscle or unit of living material or protoplasm (cell-theory of Schwann).

On the other hand, the astronomical theories of
development of the solar system from a gaseous condition to its present form, put forward by Kant and
by Laplace, had impressed men's minds with the conception of a general movement of spontaneous progress
or development in all nature; and, though such ideas

were not new but are to be found in some of the
ancient Greek philosophers, yet now for the first time
they could be considered with a sufficient knowledge
and certainty as to the facts, due to the careful observation of the two preceding centuries.   The science of
Geology came into existence, and the whole panorama
of successive stages of the earth's history, each with
its distinct population of strange animals and plants,
unlike those of the present day and simpler in proportion as they recede into the past, was revealed by Cuvier,
Agassiz, and others.   The history of the crust of the
earth was explained by Lyell as due to a process of
slow development, in order to effect which he called
in no cataclysmic agencies, no mysterious forces differing from those operating at the present day.   Thus
he carried on the narrative of orderly development
from the point at which it was left by Kant and
Laplace,—explaining by reference to the ascertained
laws of physics and chemistry the configuration of the
earth, its mountains and seas, its igneous and its
stratified rocks, just as the astronomers had explained
by those same laws the evolution of the sun and planets
from diffused gaseous matter of high temperature.

The suggestion that living things must also be
included in this great development was obvious.
They had been so included by poet-philosophers in
past ages ; they were so included by many a simple-
minded student of nature who, watching the growth
of the tree from the seed, formed a true but unverified
inference in favour of a general process of growth and
development of all things from simpler beginnings.

The delay in the establishment of the doctrine of organic evolution was due, not to the ignorant and unobservant, but to the leaders of zoological and botanical science. Knowing as they did the almost endless complexity of organic structures, realising as they did that man himself with all the mystery of his life and consciousness must be included in any explanation of the origin of living things, they preferred to regard living things as something apart from the rest of nature, specially cared for, specially created by a Divine Being, rather than to indulge in hypotheses which seemed to be beyond all possibility of proof, and were rather of the nature of poets' dreams than in accordance with the principles of that new philosophy of rigid adherence to fact and demonstration which had hitherto served as the mainsprings of scientific progress. Thus it was that the so-called " Naturphilosophen" of the last decade of the eighteenth century, and their successors in the first quarter of the nineteenth, found few adherents among the working zoologists and botanists. Lamarck, Treviranus, Erasmus Darwin, Goethe, and Saint-Hilaire preached to deaf ears, for they advanced the theory that living beings had developed by a slow process of transmutation in successive generations from simpler ancestors, and in the beginning from simplest formless matter, without being able to demonstrate any existing mechanical causes by which such development must necessarily be brought about. They were met in fact by the criticism that possibly such a development had taken place ; but, as no one could show as a simple fact of observation that it *had* taken place, nor as a result of legitimate inference that it *must* have taken place, it was quite as likely that the past and present species of animals and plants had been separately created or individually brought into existence by unknown and inscrutable causes, and (it was held) the truly scientific man would refuse to occupy himself with such fancies, whilst ever continuing to concern himself with the observation and record of indisputable facts.   The critics did well ; for the " Natur-philosophen," though right in their main conception, were premature.

It was reserved for Charles Darwin, in the year 1859, to place the whole theory of organic evolution on a new footing, and by his discovery of a mechanical cause actually existing and demonstrable, by which organic evolution must be brought about, to entirely change the attitude in regard to it of even the most rigid exponents of the scientific method.   Since its first publication in 1859 the history of Darwin's theory has been one of continuous and decisive conquest, so that at the present day it is universally accepted as the central, all-embracing doctrine of zoological and botanical science.

Darwin succeeded in establishing the doctrine of organic evolution by the introduction into the web of the zoological and botanical sciences of a new science. The subject-matter of this new science, or branch of biological science, had been neglected : it did not form part of the studies of the collector and systematist, nor was it a branch of anatomy, nor of the physiology pursued by medical men, nor again was it included in the field of microscopy and the cell-theory. The area of biological knowledge which Darwin was the first to subject to scientific method and to render, as it were, contributory to the great stream formed by the union of the various branches, the outlines of which we have already traced, is that which relates to the breeding of animals and plants, their congenital variations, and the transmission and perpetuation of those variations. This branch of biological science may be called Thremmatology ($\theta\rho\acute{\epsilon}\mu\mu\alpha$, "a thing bred"). Outside the scientific world an immense mass of observation and experiment had grown up in relation to this subject. From the earliest times the shepherd, the farmer, the horticulturist, and the "fancier" had for practical purposes made themselves acquainted with a number of biological laws, and successfully applied them without exciting more than an occasional notice from the academic students of Biology. It is one of Darwin's great merits to have made use of these observations and to have formulated their results to a large extent as the laws of variation and heredity. As the breeder selects a congenital variation which suits his requirements, and by breeding from the animals (or plants) exhibiting that variation obtains a new breed specially characterised by that variation, so in nature is there a selection amongst all the congenital variations of each generation of a species. This selection depends on the fact that more young are born than the natural provision of food will support. In consequence of this excess

of births there is a struggle for existence and a survival of the fittest, and consequently an ever-present
necessarily-acting selection, which either maintains
accurately the form of the species from generation to
generation, or leads to its modification in correspond-
ence with changes in the surrounding circumstances
which have relation to its fitness for success in the
struggle for life.

Darwin's introduction of Thremmatology into the
domain of scientific Biology was accompanied by a
new and special development of a branch of study
which had previously been known as Teleology, the
study of the adaptation of organic structures to the
service of the organisms in which they occur.   It
cannot be said that previously to Darwin there had
been any very profound study of Teleology, but it had
been the delight of a certain type of mind—that of
the lovers of nature or naturalists *par excellence*, as
they were sometimes termed—to watch the habits of
living animals and plants, and to point out the remarkable ways in which the structure of each variety
of organic life was adapted to the special circumstances of life of the variety or species.   The astonishing colours and grotesque forms of some animals and
plants which the museum zoologists gravely described
without comment were shown by these observers of
living nature to have their significance in the economy of the organism possessing them ; and a general
doctrine was recognised, to the effect that no part or
structure of an organism is without definite use and
adaptation, being *designed* by the Creator for the

benefit of the creature to which it belongs, or else for
the benefit, amusement, or instruction of his highest
creature—man. Teleology in this form of the
doctrine of design was never very deeply rooted
amongst scientific anatomists and systematists. It
was considered permissible to speculate somewhat
vaguely on the subject of the utility of this or that
startling variety of structure; but few attempts,
though some of great importance, were made to
systematically explain by observation and experiment
the adaptation of organic structures to particular purposes in the case of the lower animals and plants.
Teleology had, however, an important part in the
evelopment of what is called Physiology, viz. the
knowledge of the mechanism, the physical and
chemical properties, of the parts of the body of man
and the higher animals allied to him. The doctrine
of organs and functions—the organ designed so as to
execute the function, and the whole system of organs
and functions building up a complex mechanism, the
complete animal or plant—was teleological in origin,
and led to brilliant discoveries in the hands of the
physiologists of the last and the preceding century.
As applied to lower and more obscure forms of life,
Teleology presented almost insurmountable difficulties;
and consequently, in place of exact experiment and
demonstration, the most reckless though ingenious
assumptions were made as to the utility of the parts
and organs of lower animals, which tended to bring
so-called Comparative Physiology and Teleology generally into disrepute. Darwin's theory had as one of

its results the reformation and rehabilitation of Teleology. According to that theory, every organ, every part, colour, and peculiarity of an organism, must either be of benefit to that organism itself, or have been so to its ancestors :[1] no peculiarity of structure or general conformation, no habit or instinct in any organism, can be supposed to exist for the benefit or amusement of another organism, not even for the delectation of man himself. Necessarily, according to the theory of natural selection, structures either are present because they are selected as useful, or because they are still inherited from ancestors to whom they were useful, though no longer useful to the existing representatives of those ancestors.

The conception thus put forward entirely refounded Teleology. Structures previously inexplicable were explained as survivals from a past age, no longer useful though once of value. Every variety of form and colour was urgently and absolutely called upon to produce its title to existence either as an active useful agent or as a survival. Darwin himself spent a large part of the later years of his life in thus extending the new Teleology. A beginning only has as yet been made in the new life of that branch of zoological and botanical study.

[1] At the same time the fact that a useless variation of one part or organ may be a necessary accompaniment of a useful variation of another part or organ, was pointed out by Darwin, who applied the terms "correlation of variation," and "correlation of growth" to this phenomenon. The importance of the "solidarity" of the constituent parts of the organism can scarcely be over-estimated : it was recognised in other terms by Cuvier and the earlier morphologists.

The old doctrine of types, which was used by the philosophically-minded zoologists (and botanists) of the first half of the century as a ready means of explaining the failures and difficulties of the doctrine of design, fell into its proper place under the new dispensation. The adherence to type, the favourite conception of the transcendental morphologist, was seen to be nothing more than the expression of one of the laws of Thremmatology, viz. the persistence of hereditary transmission of ancestral characters, even when they have ceased to be significant or valuable in the struggle for existence : whilst the so-called evidences of design which was supposed to modify the limitations of types assigned to himself by the Creator were seen to be adaptations due to the selection and intensification by selective breeding of fortuitous congenital variations, which happened to prove more useful than the many thousand other variations which did not survive in the struggle for existence.

Thus not only did Darwin's theory give a new basis to the study of organic structure, but, whilst rendering the general theory of organic evolution equally acceptable and necessary, it explained the existence of low and simple forms of life as survivals of the earliest ancestry of more highly complex forms, and revealed the classifications of the systematist as unconscious attempts to construct the genealogical tree or pedigree of plants and animals. Finally, it brought the simplest living matter or formless protoplasm before the mental vision as the starting-point whence, by the operation of necessary mechanical

causes, the highest forms have been evolved, and it
rendered unavoidable the conclusion that this earliest
living material was itself evolved by gradual processes,
the result also of the known and recognised laws of
physics and chemistry, from material which we should
call not living.   It abolished the conception of life as
an entity above and beyond the common properties
of matter, and led to the conviction that the marvellous and exceptional qualities of that which we call
"living" matter are nothing more nor less than an
exceptionally complicated development of those chemical and physical properties which we recognise in a
gradually ascending scale of evolution in the carbon
compounds, containing nitrogen as well as oxygen,
sulphur, and hydrogen as constituent atoms of their
enormous molecules.   Thus mysticism was finally
banished from the domain of Biology, and Zoology
became one of the physical sciences,—the science
which seeks to arrange and discuss the phenomena of
animal life and form as the outcome of the operation
of the laws of physics and chemistry.

Nature and Scope of Zoology

The brief historical outline above given is sufficient
to justify us in rejecting, for the purposes of an adequate appreciation of the history and scope of Zoology,
that simple division of the science into Morphology
and Physiology which is a favourite one at the present
day.   No doubt the division is a logical one, based as
it is upon the distinction of the study of form and

structure in themselves (Morphology) from the study of what are the activities and functions of the forms and structures (Physiology). Such logical divisions are possible upon a variety of bases, but are not necessarily conducive to the ascertainment and remembrance of the historical progress and present significance of the science to which they are applied. As a matter of convenience and as the outcome of historical events it happens that in the universities of Europe, whilst Botany in its entirety is usually represented by one chair, the animal side of Biology is represented by a chair of so-called Zoology, which is understood as the old-fashioned systematic Zoology, a chair of human and comparative Anatomy, and a chair of Physiology (signifying the mechanics, physics, and chemistry of animals especially in relation to man). Fifty years ago the chairs of Anatomy and Physiology were united in one. No such distinction of mental activities as that involved in the division of the study of animal life into Morphology and Physiology has ever really existed : the investigator of animal forms has never entirely ignored the functions of the forms studied by him, and the experimental inquirer into the functions and properties of animal tissues and organs has always taken very careful account of the forms of those tissues and organs.

A more instructive subdivision of the science of animal Biology or Zoology is one which shall correspond to the separate currents of thought and mental preoccupation which have been historically manifested in western Europe in the gradual evolution of what is to-day the great river of zoological doctrine to which they have all been rendered contributory. Such a subdivision of Zoology, whilst it enables us to trace the history of thought, corresponds very closely with the actual varieties of mental attitude exhibited at the present day by the devotees of zoological study, though it must be remembered that the gathering together of all the separate currents by Darwin is certain sooner or later to entail new developments and branchings of the stream.

We accordingly recognise the following five branches of zoological study :—

1. *Morphography.*—The work of the collector and systematist: exemplified by Linnæus and his predecessors, by Cuvier, Agassiz, Haeckel.

2. *Bionomics.*—The lore of the farmer, gardener. sportsman, fancier, and field-naturalist, including Thremmatology, or the science of breeding, and the allied Teleology, or science of organic adaptations : exemplified by the patriarch Jacob, the poet Virgil, Sprengel, Kirby and Spence, Wallace and Darwin.

3. *Zoo-Dynamics, Zoo-Physics, Zoo-Chemistry.* —The pursuit of the learned physician,— Anatomy and Physiology: exemplified by Harvey, Haller, Hunter, Johann Müller and the modern school of experimental physiologists.

4. *Plasmology.*—The study of the ultimate corpuscles of living matter, their structure, development, and properties, by the aid of the

microscope ; exemplified by Malpighi, Hook, Schwann, Kowalewsky, and Metschnikoff.

5. *Philosophical Zoology.*—General conceptions with regard to the relations of living things (especially animals) to the universe, to man, and to the Creator, their origin and significance : exemplified in the writings of the philosophers of classical antiquity, and of Linnæus, Goethe, Lamarck, Cuvier, Lyell, H. Spencer, and Darwin.

It is true that it is impossible to assign the great names of the present century to a single one of the subdivisions of the science thus recognised. With men of an earlier date such special assignment is possible, and there would be no difficulty about thus separating the minor specialists of modern times. But the fact is that as we approach Darwin's epoch we find the separate streams more and more freely connected with one another by anastomosing branches. The men who have left their mark on the progress of science have been precisely those who have been instrumental in bringing about such confluence, and have distinguished themselves by the influence of their discoveries or generalisations upon several lines of work. At last, in Darwin we find a name which might appear in each of our subdivisions,—a zoologist to whose doctrine all are contributory, and by whose labours all are united and reformed.

We shall now briefly sketch the history of these streams of thought, premising that one has (so far as the last three centuries are concerned) but little start

of another, and that sooner or later the influence of
the progress of one branch makes itself felt in the
progress of another.

MORPHOGRAPHY

Under this head we include the systematic exploration and tabulation of the facts, involved in the
recognition of all the recent and extinct kinds of
animals and their distribution in space and time.
The chief varieties of zoological workers coming under
this head are (1) the museum-makers of old days and
their modern representatives the curators and de-
scribers of zoological collections, (2) early explorers
and modern naturalist-travellers and writers on zoo-
geography, and (3) collectors of fossils and palæonto-
logists.   Gradually since the time of Hunter and
Cuvier anatomical study has associated itself with the
more superficial morphography until to-day no one
considers a study of animal form of any value which
does not include internal structure, Histology, and
Embryology in its scope.

The real dawn of Zoology after the legendary
period of the Middle Ages is connected with the name
of an Englishman, Wotton, born at Oxford in 1492,
who practised as a physician in London and died in
1555.   He published a treatise *De Differentiis Animalium* at Paris in 1552.   In many respects Wotton
was simply an exponent of Aristotle, whose teaching,
with various fanciful additions, constituted the real
basis of zoological knowledge throughout the Middle
Ages.   It was Wotton's merit that he rejected the

legendary and fantastic accretions, and returned to
Aristotle and the observation of nature.   The most
ready means of noting the progress of Zoology during
the sixteenth, seventeenth, and eighteenth centuries is
to compare the classificatory conceptions of successive
naturalists with those which are to be found in the works
of Aristotle himself.   Aristotle did not definitely and
in tabular form propound a classification of animals,
but from a study of his treatises *Historia Animalium*,
De Generatione Animalium, and *De Partibus Anima-*
lium the following classification can be arrived at :—

A. Ἔναιμα, blood-holding animals (= *Vertebrata*).
1. Ζωοτοκοῦντα ἐν αὑτοῖς, viviparous *Enæma* (= Mammals, including the Whale).
2. Ὄρνιθες (= Birds).
3. Τετράποδα ἤ ἄποδα ὠοτοκοῦντα, four-footed or legless *Enæma*
which lay eggs (= Reptiles and *Amphibia*).
4. Ἰχθύες (= Fishes).
B. Ἄναιμα, bloodless animals (= *Invertebrata*).
1. Μαλάκια, soft-bodied *Anæma* (= *Cephalopoda*).
2. Μαλακόστρακα, soft-shelled *Anæma* (= *Crustacea*).
3. Ἔντομα, insected *Anæma* or Insects (= *Arthropoda*, exclusive of *Crustacea*).
4. Ὀστρακοδέρμματα, shell-bearing *Anæma* (= *Echini*, *Gastro-*
poda, and *Lamellibranchia*).

Wotton follows Aristotle in the division of animals
into the *Enæma* and the *Anæma*, and in fact in the
recognition of all the groups above given, adding only
one large group to those recognised by Aristotle under
the *Anæma*, namely, the group of *Zoophyta*, in which
Wotton includes the Holothuriæ, Star-Fishes, Medusæ,
Sea-Anemones, and Sponges.   Wotton divides the
viviparous quadrupeds into the many-toed, double-
hoofed, and single-hoofed.   By the introduction of a

method of classification which was due to the superficial Pliny,—viz. one depending, not on structure,
but on the medium inhabited by an animal, whether
earth, air, or water,—Wotton is led to associate Fishes
and Whales as aquatic animals.   But this is only a
momentary lapse, for he broadly distinguishes the two
kinds.

Conrad Gesner (1516-1565), who was a physician
and held professorial chairs in various Swiss cities, is
the most voluminous and instructive of these earliest
writers on systematic Zoology, and was so highly
esteemed that his *Historia Animalium* was repub-
lished a hundred years after his death.   His great
work appeared in successive parts,—*e.g. Vivipara,
Ovipara, Aves, Pisces, Serpentes et Scorpio,*—and contains descriptions and illustrations of a large number
of animal forms with reference to the lands inhabited
by them.   Gesner's work, like that of John Johnstone
(*b.* 1603), who was of Scottish descent and studied at
St. Andrews, and like that of Ulysses Aldrovandi of
Bologna (*b.* 1522), was essentially a compilation, more
or less critical, of all such records, pictures, and
relations concerning beasts, birds, reptiles, fishes, and
monsters as could be gathered together by one reading
in the great libraries of Europe, travelling from city
to city, and frequenting the company of those who
either had themselves passed into distant lands or
possessed the letters written and sometimes the specimens brought home by adventurous persons.

The exploration of parts of the New World next
brought to hand descriptions and specimens of many

novel forms of animal life, and in the latter part of the sixteenth century and the beginning of the seventeenth that careful study by " specialists " of the structure and life-history of particular groups of animals was commenced which, directed at first to common and familiar kinds, was gradually extended until it formed a sufficient body of knowledge to serve as an anatomical basis for classification.   This minuter study had two origins, one in the researches of the medical anatomists, such as Fabricius (1537-1619), Severinus (1580-1656), Harvey (1578-1657), and Tyson (1649-1708), the other in the careful work of the entomologists and first microscopists, such as Malpighi (1628-1694), Swammerdam (1637-1680), and Hook (1635-1702). The commencement of anatomical investigations deserves notice here as influencing the general accuracy and minuteness with which zoological work was prosecuted, but it was not until a late date that their full influence was brought to bear upon systematic Zoology by Georges Cuvier (1769-1832).

The most prominent name between that of Gesner and Linnæus in the history of systematic Zoology is that of John Ray.   Though not so extensive as that of Linnæus, his work is of the highest importance, and rendered the subsequent labours of the Swedish naturalist far easier than they would otherwise have been.   A chief merit of Ray is to have limited the term " species " and to have assigned to it the significance which it has until the Darwinian era borne, whereas previously it was loosely and vaguely applied. He also made considerable use of anatomical characters in his definitions of larger groups, and may thus be considered as the father of modern Zoology. Associated with Ray in his work, and more especially occupied with the study of the Worms and Mollusca, was Martin Lister (1638-1712), who is celebrated also as the author of the first geological map.

After Ray's death in London in 1705 the progress of anatomical knowledge, and of the discovery and illustration of new forms of animal life from distant lands, continued with increasing vigour. We note the names of Vallisnieri (1661-1730) and Alexander Monro (1697-1767); the travellers Tournefort (1656-1708) and Shaw (1692-1751); the collectors Rumphius (1637-1706) and Hans Sloane (1660-1753); the entomologist Réaumur (1683-1757); Lhwyd (1703) and Linck (1674-1734), the students of Star-Fishes; Peyssonel (b. 1694) the investigator of Polyps and the opponent of Marsigli and Réaumur, who held them to be plants; Woodward, the palæontologist (1665-1722),—not to speak of others of less importance.

Two years after Ray's death Carl Linnæus was born. Unlike Jacob Theodore Klein, the town-clerk of Dantzig (1685-1759), whose careful treatises on various groups of plants and animals were published during the period between Ray and Linnæus, the latter had his career marked out for him in a university, that of Upsala, where he was first professor of medicine and subsequently of natural history. His lectures formed a new departure in the academic treatment of Zoology and Botany, which, in direct continuity from the Middle Ages, had hitherto been

subjected to the traditions of the medical profession
and regarded as mere branches of "materia medica."
Linnæus taught Zoology and Botany as branches of
knowledge to be studied for their own intrinsic
interest, and not for the sake of the "simples" yielded
by animals and plants to the pharmacopœia. His
great work, the *Systema Naturæ*, ran through twelve
editions during his lifetime (1st ed. 1735, 12th 1768).
Apart from his special discoveries in the anatomy of
plants and animals, and his descriptions of new species,
the great merit of Linnæus was his introduction of a
method of enumeration and classification which may
be said to have created systematic Zoology and Botany
in their present form, and established his name for
ever as the great organiser, the man who recognised
a great practical want in the use of language and
supplied it. Linnæus adopted Ray's conception of
species, but he made species a practical reality by
insisting that every species shall have a double Latin
name,—the first half to be the name of the genus
common to several species, and the second half to be
the specific name. Previously to Linnæus long many-
worded names had been used, sometimes with one
additional adjective, sometimes with another, so that
no true *names* were fixed and accepted. Linnæus by
his binomial system made it possible to write and
speak with accuracy of any given species of plant or
animal. He was, in fact, the Adam of zoological
science. He proceeded further to introduce into his
enumeration of animals and plants a series of groups,
viz. genus, order, class, which he compared to the

subdivisions of an army or the subdivisions of a
territory, the greater containing several of the less, as
follows :—

Class.	Order.	Genus.	Species.	Variety.
Genus sum- mum.	Genus inter- medium.	Genus proxi- mum.	Species.	Individuum.
Provincia.	Territorium.	Parœcia.	Pagus.	Domicilium.
Legio.	Cohors.	Manipulus.	Contubernium.	Miles.

Linnæus himself recognised the purely subjective
character of his larger groups ; for him species were,
however, objective : "There are," he said, "just so
many species as in the beginning the Infinite Being
created." It was reserved for a philosophic zoologist
of the nineteenth century (Agassiz, *Essay on Classification*, 1859) to maintain dogmatically that genus, order,
and class were also objective facts capable of precise
estimation and valuation. This climax was reached at
the very moment when Darwin was publishing the
Origin of Species (1859), by which universal opinion
has been brought to the position that species, as well
as genera, orders, and classes, are the subjective expressions of a vast ramifying pedigree in which the
only objective existences are individuals, the apparent species as well as higher groups being marked
out, not by any distributive law, but by the purely
non-significant operation of human experience, which
cannot transcend the results of death and decay.

The classification of Linnæus (from *Syst. Nat.*
12th ed. 1766) should be compared with that of Aristotle. It is as follows,—the complete list of Linnæan
genera being here reproduced :—

Class I. MAMMALIA.

    Order 1. *Primates.*
        Genera : *Homo, Simia, Lemur, Vespertilio.*
      „   2. *Bruta.*
        Genera : *Elephas, Trichecus, Bradypus, Myrme-
        cophaga, Manis, Dasypus.*
      „   3. *Feræ.*
        Genera : *Phoca, Canis, Felis, Viverra, Mustela,
        Ursus, Didelphys, Talpa, Sorex, Erinaceus.*
      „   4. *Glires.*
        Genera : *Hystrix, Lepus, Castor, Mus, Sciurus,
        Noctilio.*
      „   5. *Pecora.*
        Genera : *Camelus, Moschus, Cervus, Capra, Ovis,
        Bos.*
      „   6. *Belluæ.*
        Genera : *Equus, Hippopotamus, Sus, Rhinoceros.*
      „   7. *Cete.*
        Genera : *Monodon, Balæna, Physeter, Delphinus.*

Class II. AVES.

    Order 1. *Accipitres.*
        Genera : *Vultur, Falco, Strix, Lanius.*
      „   2. *Picæ.*
        Genera : (a) *Trochilus, Certhia, Upupa, Buphaga,
        Sitta, Oriolus, Coracias, Gracula, Corvus, Para-
        disea ;* (b) *Ramphastos, Trogon, Psittacus, Cro-
        tophaga, Picus, Yunx, Cuculus, Bucco ;* (c)
        Buceros, Alcedo, Merops, Todos.
      „   3. *Anseres.*
        Genera : (a) *Anas, Mergus, Phæthon, Plotus ;* (b)
        *Rhyncops, Diomedea, Alca, Procellaria, Pele-
        canus, Larus, Sterna, Colymbus.*
      „   4. *Grallæ.*
        Genera : (a) *Phœnicopterus, Platalea, Palamedea,
        Mycteria, Tantalus, Ardea, Recurvirostra, Scolo-
        pax, Tringa, Fulica, Parra, Rallus, Psophia,
        Cancroma ;* (b) *Hematopus, Charadrius, Otis,
        Struthio.*
      „   5. *Gallinæ.*
        Genera : *Didus, Pavo, Meleagris, Crax, Phasi-
        anus, Tetrao, Numida.*

Order 6. *Passeres.*

> Genera : (*a*) *Loxia, Fringilla, Emberiza;* (*b*) *Caprimulgus, Hirundo, Pipra ;* (*c*) *Turdus, Ampelis, Tanagra, Muscicapa ;* (*d*) *Parus, Motacilla, Alauda, Sturnus, Columba.*

Class III. AMPHIBIA.

Order 1. *Reptilia.*

> Genera : *Testudo, Draco, Lacerta, Rana.*

„ 2. *Serpentes.*

> Genera : *Crotalus, Boa, Coluber, Anguis, Amphisbæna, Cæcilia.*

„ 3. *Nantes.*

> Genera : *Petromyzon, Raja, Squalus, Chimæra, Lophius, Acipenser, Cyclopterus, Balistes, Ostracion, Tetrodon, Diodon, Centriscus, Syngnathus, Pegasus.*

Class IV. PISCES.

Order 1. *Apodes.*

> Genera : *Muræna, Gymnotus, Trichiurus, Anarrhichas, Ammodytes, Ophidium, Stromateus, Xiphias.*

„ 2. *Jugulares.*

> Genera : *Callionymus, Uranoscopus, Trachinus, Gadus, Blennius.*

„ 3. *Thoracici.*

> Genera : *Cepola, Echeneis, Coryphæna, Gobius, Cottus, Scorpæna, Zeus, Pleuronectes, Chætodon, Sparus, Labrus, Sciæna, Perca, Gasterosteus, Scomber, Mullus, Trigla.*

„ 4. *Abdominales.*

> Genera : *Cobitis, Amia, Silurus, Zeuthis, Loricaria, Salmo, Fistularia, Esox, Elops, Argentina, Atherina, Mugil, Mormyrus, Exocætus, Polynemus, Clupea, Cyprinus.*

Class V. INSECTA.

Order 1. *Coleoptera.*

> Genera : (*a*) *Scarabæus, Lucanus, Dermestes, Hister, Byrrhus, Gyrinus, Attelabus, Curculio, Silpha, Coccinella ;* (*b*) *Bruchus, Cassida, Ptinus, Chrysomela, Hispa, Meloe, Tenebrio, Lampyris, Mordella, Staphylinus ;* (*c*) *Ceram-*

byx, *Leptura, Cantharis, Elater, Cicindela, Bu-*
prestis, Dytiscus, Carabus, Necydalis, Forficula.

Order 2. *Hemiptera.*

Genera: *Blatta, Mantis, Gryllus, Fulgora, Ci-*
cada, Notonecta, Nepa, Cimex, Aphis, Chermes,
Coccus, Thrips.

„   3. *Lepidoptera.*

Genera: *Papilio, Sphinx, Phalæna.*

„   4. *Neuroptera.*

Genera: *Libellula, Ephemera, Myrmeleon, Phry-*
ganea, Hemerobius, Panorpa, Raphidia.

„   5. *Hymenoptera.*

Genera: *Cynips, Tenthredo, Sirex, Ichneumon,*
Sphex, Chrysis, Vespa, Apis, Formica, Mutilla.

„   6. *Diptera.*

Genera: *Œstrus, Tipula, Musca, Tabanus, Culex,*
Empis, Conops, Asilus, Bombylius, Hippobosca.

„   7. *Aptera.*

Genera: (*a*) Pedibus sex; capite a thorace discreto: *Lepisma, Podura, Termes, Pedicu-*
lus, Pulex.

(*b*) Pedibus 8-14; capite thoraceque unitis:
Acarus, Phalangium, Aranea, Scorpio,
Cancer, Monoculus, Oniscus.

(*c*) Pedibus pluribus; capite a thorace discreto:
Scolopendra, Julus.

Class VI. Vermes.

Order 1. *Intestina.*

Genera: (*a*) Pertusa laterali poro: *Lumbricus,*
Sipunculus, Fasciola.

(*b*) Imperforata: poro laterali nullo: *Gordius,*
Ascaris, Hirudo, Myxine.

„   2. *Mollusca.*

Genera: (*a*) Ore supero; basi se affigens: *Ac-*
tinia, Ascidia.

(*b*) Ore antico; corpore pertuso laterali fora-
minulo: *Limax, Aplysia, Doris, Tethis.*

(*c*) Ore antico; corpore tentaculis antice cincto:
Holothuria, Terebella.

(*d*) Ore antico; corpore brachiato: *Triton,*
Sepia, Clio, Lernæa, Scyllæa.

(e) Ore antico; corpore pedato: *Aphrodita*, *Nereis*.

(f) Ore infero centrali: *Medusa*, *Asteria*, *Echinus*.

Order 3. *Testacea*.

Genera : (a) Multivalvia : *Chiton*, *Lepas*, *Pholas.*

(b) Bivalvia (= *Conchæ*): *Mya*, *Solen*, *Tellina*, *Cardium*, *Mactra*, *Donax*, *Venus*, *Spondylus*, *Chama*, *Arca*, *Ostrea*, *Anomia*, *Mytilus*, *Pinna*.

(c) Univalvia spira regulari (= *Cochleæ*): *Argonauta*, *Nautilus*, *Conus*, *Cypræa*, *Bulla*, *Voluta*, *Buccinum*, *Strombus*, *Murex*, *Trochus*, *Turbo*, *Helix*, *Nerita*, *Haliotis*.

(d) Univalvia absque spira regulari : *Patella*, *Dentalium*, *Serpula*, *Teredo*, *Sabella*.

„ 4. *Lithophyta*.

Genera : *Tubipora*, *Madrepora*, *Millepora*, *Cellepora*.

„ 5. *Zoophyta*.

Genera : (a) Fixata : *Isis*, *Gorgonia*, *Alcyonium*, *Spongia*, *Flustra*, *Tubularia*, *Corallina*, *Sertularia*, *Vorticella*.

(b) Locomotiva : *Hydra*, *Pennatula*, *Tænia*, *Volvox*, *Furia*, *Chaos*.

The characters of the six classes are thus given by Linnæus :—

Cor biloculare, biauritum ;	viviparis, *Mammalibus;*
Sanguine calido, rubro :	oviparis, *Avibus.*
Cor uniloculare, uniauritum;[1]	pulmone arbitrario, *Amphibiis;*
Sanguine frigido, rubro :	branchiis externis, *Piscibus.*
Cor uniloculare, inauritum ;	antennatis, *Insectis;*
Sanie frigida, albida :	tentaculatis, *Vermibus.*

---

[1] The anatomical error in reference to the auricles of Reptiles and Batrachians on the part of Linnæus is extremely interesting, since it shows to what an extent the most patent facts may escape the observation of even the greatest observers, and what an amount of repeated dissection and unprejudiced attention has been necessary before the structure of the commonest animals has become known.

Between Linnæus and Cuvier there are no very
great names; but under the stimulus given by the
admirable method and system of Linnæus observation
and description of new forms from all parts of the
world, both recent and fossil, accumulated. We can
only cite the names of Charles Bonnet (1720-1793),
the entomologist, who described the reproduction of
Aphis; Banks and Solander, who accompanied Captain Cook on his first voyage (1768-1771); Thomas
Pennant (1726-1798), the describer of the English
fauna; Peter Simon Pallas (1741-1811), who specially
extended the knowledge of the Linnæan *Vermes,* and
under the patronage of the empress Catherine explored
Russia and Siberia; De Geer (1720-1778), the entomologist; Lyonnet (1707-1789), the author of the
monograph of the anatomy of the caterpillar of *Cossus
ligniperdus;* Cavolini (1756-1810), the Neapolitan
marine zoologist and forerunner of Della Chiaje (fl.
1828); O. F. Müller (1730-1784), the describer of
freshwater *Oligochæta;* Abraham Trembley (1700-
1784), the student of *Hydra;* and Ledermüller (1719-
1769), the inventor of the term *Infusoria.* The effect
of the Linnæan system upon the general conceptions
of zoologists was no less marked than were its results
in the way of stimulating the accumulation of accurately observed details. The notion of a *scala naturæ,*
which had since the days of classical antiquity been a
part of the general philosophy of nature amongst those
who occupied themselves with such conceptions, now
took a more definite form in the minds of skilled
zoologists. The species of Linnæus were supposed to

represent a series of steps in a scale of ascending complexity, and it was thought possible thus to arrange the animal kingdom in a single series,—the orders within the classes succeeding one another in regular gradation, and the classes succeeding one another in a similar rectilinear progression.   Lamarck represents most completely, both by his development theory (to be further mentioned below) and by his scheme of classification, the high-water mark of the popular but fallacious conception of a *scala naturæ*.   His classification (1801-1812) is as follows :—

Invertebrata.
> 1. *Apathetic Animals.*
>> Class I. INFUSORIA.
>>> Orders : *Nuda, Appendiculata.*
>> Class II. POLYPI.
>>> Orders : *Ciliati* (*Rotifera*), *Denudati* (Hydroids), *Vaginati* (*Anthozoa* and *Polyzoa*), *Natantes* (Crinoids).
>> Class III. RADIARIA.
>>> Orders : *Mollia* (*Acalephæ*), *Echinoderma* (including *Actiniæ*).
>> Class IV. TUNICATA.
>>> Orders : *Bothryllaria, Ascidia.*
>> Class V. VERMES.
>>> Orders : *Molles* (Tape-Worms and Flukes), *Rigiduli* (Nematoids), *Hispiduli* (Nais, etc.), *Epizoariæ* (Lernæans, etc.)
> 2. *Sensitive Animals.*
>> Class VI. INSECTA (*Hexapoda*).
>>> Orders : *Aptera, Diptera, Hemiptera, Lepidoptera, Hymenoptera, Neuroptera, Orthoptera, Coleoptera.*
>> Class VII. ARACHNIDA.
>>> Orders : *Antennato-Trachealia* (= *Thysanura* and *Myriapoda*), *Exantennato-Trachealia, Exantennato-Branchialia.*
>> Class VIII. CRUSTACEA.
>>> Orders : *Heterobranchia* (*Branchiopoda, Isopoda, Amphipoda, Stomapoda*), *Homobranchia* (*Decapoda*).

Class IX. ANNELIDA.
  Orders: *Apoda, Antennata, Sedentaria.*
Class X. CIRRIPEDIA.
  Orders: *Sessilia, Pedunculata.*
Class XI. CONCHIFERA.
  Orders: *Dimyaria, Monomyaria.*
Class XII. MOLLUSCA.
  Orders: *Pteropoda, Gasteropoda, Trachelipoda, Cephalopoda, Heteropoda.*

Vertebrata.
  3. *Intelligent Animals.*
    Class XIII. FISHES.   Class XV. BIRDS.
      „ XIV. REPTILES.   „ XVI. MAMMALS.

The enumeration of orders above given will enable the reader to form some conception of the progress of knowledge relating to the lower forms of life during the fifty years which intervened between Linnæus and Lamarck. The number of genera recognised by Lamarck is more than ten times as great as that recorded by Linnæus.

We have mentioned Lamarck before his great contemporary Cuvier because, in spite of his valuable philosophical doctrine of development, he was, as compared with Cuvier and estimated as a systematic zoologist, a mere enlargement and logical outcome of Linnæus.

The distinctive merit of Cuvier is that he started a new view as to the relationships of animals, which he may be said in a large measure to have demonstrated as true by his own anatomical researches. He opposed the *scala naturæ* theory, and recognised four distinct and divergent branches or *embranchemens*, as he called them, in each of which he arranged a certain number of the Linnæan classes, or similar classes. The *embranchemens* were characterised each by a dif-

ferent type of anatomical structure. Cuvier thus laid
the foundation of that branching tree-like arrangement
of the classes and orders of animals which we now
recognise as being the necessary result of attempts to
represent what is practically a genealogical tree or
pedigree. Apart from this, Cuvier was a keen-sighted
and enthusiastic anatomist of great skill and industry.
It is astonishing how many good observers it requires
to dissect and draw and record over and over again
the structure of an animal before an approximately
correct account of it is obtained. Cuvier dissected
many Molluscs and other animals which had not previously been anatomised; of others he gave more
correct accounts than had been given by earlier
writers. Skilful as he was, his observations are very
frequently erroneous. Great accuracy in work as well
as great abundance of production has only distinguished one amongst all the great names of Zoology
—that of Johann Müller. It certainly did not distinguish Cuvier. Another special distinction of Cuvier
is his remarkable work in comparing extinct with
recent organisms, his descriptions of the fossil *Mammalia* of the Paris basin, and his general application
of the knowledge of recent animals to the reconstruction of extinct ones, as indicated by fragments only
of their skeletons.

It was in 1812 that Cuvier communicated to the
Academy of Sciences of Paris his views on the classification of animals. He says—

" Si l'on considère le règne animal d'après les principes que nous
venons de poser, en se débarrassant des préjugés établis sur les

divisions anciennement admises, en n'ayant égard qu'à l'organisation et à la nature des animaux, et non pas à leur grandeur, à leur
utilité, au plus ou moins de connaissance que nous en avons, ni à
toutes les autres circonstances accessoires, on trouvera qu'il existe
quatre formes principales, quatre plans généraux, si l'on peut s'ex-
primer ainsi, d'après lesquels tous les animaux semblent avoir été
modelés et dont les divisions ultérieures, de quelque titre que les
naturalistes les aient décorées, ne sont que des modifications assez
légères, fondées sur le développement, ou l'addition de quelques
parties qui ne changent rien à l'essence du plan."

His classification as finally elaborated in *Le Règne
Animal* (Paris, 1829) is as follows :—

First Branch. **Animalia Vertebrata.**

    Class I. MAMMALIA.

        Orders : *Bimana, Quadrumana, Carnivora, Marsupialia,
Rodentia, Edentata, Pachydermata, Ruminantia, Cetacea.*

    Class II. BIRDS.

        Orders : *Accipitres, Passeres, Scansores, Gallinæ, Grallæ,
Palmipedes.*

    Class III. REPTILIA.

        Orders : *Chelonia, Sauria, Ophidia, Batrachia.*

    Class IV. FISHES.

        Orders : (*a*) *Acanthopterygii, Abdominales, Subbrachii,
Apodes, Lophobranchii, Plectognathi ;* (*b*) *Sturiones, Sel-
achii, Cyclostomi.*

Second Branch. **Animalia Mollusca.**

    Class I. CEPHALOPODA.

    Class II. PTEROPODA.

    Class III. GASTROPODA.

        Orders : *Pulmonata, Nudibranchia, Inferobranchia, Tecti-
branchia, Heteropoda, Pectinibranchia, Tubulibranchia,
Scutibranchia, Cyclobranchia.*

    Class IV. ACEPHALA.

        Orders : *Testacea, Tunicata.*

    Class V. BRACHIOPODA.

    Class VI. CIRRHOPODA.

Third Branch. **Animalia Articulata.**

    Class I. ANNELIDES.

        Orders : *Tubicolæ, Dorsibranchiæ, Abranchiæ.*

Class II. CRUSTACEA.
    Orders: (*a*) Malacostraca: *Decapoda, Stomapoda, Amphipoda, Lœmodipoda, Isopoda;* (*b*) Entomostraca: *Branchiopoda, Pœcilopoda, Trilobitœ.*

Class III. ARACHNIDES.
    Orders: *Pulmonariœ, Trachcariœ.*

Class IV. INSECTS.
    Orders: *Myriapoda, Thysanura, Parasita, Suctoria, Coleoptera, Orthoptera, Hemiptera, Neuroptera, Hymenoptera, Lepidoptera, Rhipiptera, Diptera.*

Fourth Branch. **Animalia Radiata.**

Class I. ECHINODERMS.
    Orders: *Pedicellata, Apoda.*

Class II. INTESTINAL WORMS.
    Orders: *Nematoidea, Parenchymatosa.*

Class III. ACALEPHÆ.
    Orders: *Simplices, Hydrostaticœ.*

Class IV. POLYPI (including the *Cœlentera* of later authorities and the *Polyzoa*).
    Orders: *Carnosi, Gelatinosi, Polypiarii.*

Class V. INFUSORIA.
    Orders: *Rotifera, Homogenea* (this includes the *Protozoa* of recent writers and some *Protophyta*).

The leading idea of Cuvier, his four *embranchemens*, was confirmed by the Russo-German naturalist Von Baer (1792–1876), who adopted Cuvier's divisions, speaking of them as the peripheric, the longitudinal, the massive, and the vertebrate types of structure. Von Baer, however, has another place in the history of Zoology, being the first and most striking figure in the introduction of Embryology into the consideration of the relations of animals to one another.

Cuvier may be regarded as *the* zoologist by whom anatomy was made the one important guide to the understanding of the relations of animals. But it should be noted that the belief, dating from Malpighi

(1670), that there *is* a relationship to be discovered, and not merely a haphazard congregation of varieties of structure to be classified, had previously gained ground. Cuvier was familiar with the speculations of the " Natur-philosophen," and with the doctrine of transmutation and filiation by which they endeavoured to account for existing animal forms. The noble aim of F. W. J. Schelling, " das ganze System der Naturlehre von dem Gesetze der Schwere bis zu den Bildungstrieben der Organismus als ein organisches Ganze darzustellen," which has ultimately been realised through Darwin, was a general one among the scientific men of the year 1800. Lamarck accepted the development theory fully, and pushed his speculations far beyond the realm of fact. The more cautious Cuvier adopted a view of the relationships of animals which, whilst denying genetic connection as the explanation, recognised an essential identity of structure throughout whole groups of animals. This identity was held to be due to an ultimate law of nature or the Creator's plan. The tracing out of this identity in diversity, whether regarded as evidence of blood-relationship or as a remarkable display of skill on the part of the Creator in varying the details whilst retaining the essential, became at this period a special pursuit, to which Goethe, the poet, who himself contributed importantly to it, gave the name " Morphology." C. F. Wolff, Goethe, and Oken share the credit of having initiated these views, in regard especially to the structure of flowering plants and the Vertebrate skull. Cuvier's doctrine of four plans of structure was essentially a morphological one, and so was the single-scale doctrine of Buffon and Lamarck, to which it was opposed.   Cuvier's morphological doctrine received its fullest development in the principle of the " correlation of parts," which he applied to palæontological investigation, namely, that every animal is a definite whole, and that no part can be varied without entailing correlated and law-abiding variations in other parts, so that from a fragment it should be possible, had we a full knowledge of the laws of animal structure or Morphology, to reconstruct the whole.   Here Cuvier was imperfectly formulating, without recognising the real physical basis of the phenomena, the results of the laws of heredity and variation, which were subsequently investigated and brought to bear on the problems of animal structure by Darwin.

Richard Owen [1] may be regarded as the foremost of Cuvier's disciples.   Owen not only occupied himself with the dissection of rare animals, such as the Pearly Nautilus, Lingula, Limulus, Protopterus, Apteryx, etc., and with the description and reconstruction of extinct Reptiles, Birds, and Mammals,— following the Cuvierian tradition,—but gave precision and currency to the morphological doctrines which had taken their rise in the beginning of the century by the introduction of two terms, " Homology " and " Analogy," which were defined so as to express two

[1] Born in 1804 in Lancaster ; conservator of the Hunterian Museum, London, 1830-1856 ; superintendent Nat. Hist. Brit. Mus. 1856-1884.

different kinds of agreement in animal structures, which, owing to the want of such "counters of thought," had been hitherto continually confused. *Analogous* structures in any two animals compared were by Owen defined as structures performing similar functions, but not necessarily derived from the modification of one and the same part in the "plan" or "archetype" according to which the two animals compared were supposed to be constructed. *Homologous* structures were such as, though greatly differing in appearance and detail from one another, and though performing widely different functions, yet were capable of being shown by adequate study of a series of intermediate forms to be derived from one and the same part or organ of the "plan-form" or "archetype." It is not easy to exaggerate the service rendered by Owen to the study of Zoology by the introduction of this apparently small piece of verbal mechanism; it takes place with the classificatory terms of Linnæus. And, though the conceptions of "archetypal Morphology," to which it had reference, are now abandoned in favour of a genetic Morphology, yet we should remember, in estimating the value of this and of other speculations which have given place to new views in the history of science, the words of the great reformer himself. "Erroneous observations are in the highest degree injurious to the progress of science, since they often persist for a long time. But erroneous theories, when they are supported by facts, do little harm, since every one takes a healthy pleasure in proving their falsity" (Darwin). Owen's definition of analogous structures holds good at the present day. His homologous structures are now spoken of as "homogenetic" structures, the idea of community of representation in an archetype giving place to community of derivation from a single representative structure present in a common ancestor. Darwinian Morphology has further rendered necessary the introduction of the terms "homoplasy" and "homoplastic"[1] to express that close agreement in *form* which may be attained in the course of evolutional changes by organs or parts in two animals which have been subjected to similar moulding conditions of the environment, but have no genetic community of origin, to account for their close similarity in form and structure.

The classification adopted by Owen in his lectures (1855) does not adequately illustrate the progress of zoological knowledge between Cuvier's death and that date, but, such as it is, it is worth citing here.

Province: **Vertebrata** (*Myelencephala*, Owen).
    Classes : MAMMALIA, AVES, REPTILIA, PISCES.
Province: **Articulata.**
    Classes : ARACHNIDA, INSECTA (including Sub-Classes *Myriapoda, Hexapoda*), CRUSTACEA (including Sub-Classes *Entomostraca, Malacostraca*), EPIZOA (Epizootic *Crustacea*), ANNELLATA (Chætopods and Leeches), CIRRIPEDIA.
Province: **Mollusca.**
    Classes : CEPHALOPODA, GASTEROPODA, PTEROPODA, LAMELLIBRANCHIATA, BRACHIOPODA, TUNICATA.

---

[1] See Lankester, "On the Use of the Term Homology in Modern Zoology and the Distinction between Homogenetic and Homoplastic Agreements," *Ann. and Mag. Nat. Hist.* 1870.

Province: **Radiata.**

   Sub-Province: **Radiaria.**

      Classes: ECHINODERMATA, BRYOZOA, ANTHOZOA, ACA-
      LEPHÆ, HYDROZOA.

   Sub-Province: **Entozoa.**

      Classes: CŒLELMINTHA, STERELMINTHA.

   Sub-Province: **Infusoria.**

      Classes: ROTIFERA, POLYGASTRIA (the *Protozoa* of recent
      authors).

The real centre of progress of systematic Zoology was no longer in France nor with the disciples of Cuvier in England, but after his death moved to Germany. The wave of morphological speculation, with its outcome of new systems and new theories of classification, which were as numerous as the professors of zoological science,[1] was necessarily succeeded in the true progress of the science by a period of minuter study in which the microscope, the discovery of embryological histories, and the all-important cell-theory came to swell the stream of exact knowledge.

We have already mentioned Von Baer in this connection, and given a passing reference to Johann Müller, the greatest of all investigators of animal structure in the present century. Müller (1801-1858) was in Germany the successor of Rathke (1793-1860) and of Meckel (1781-1833) as the leader of anatomical investigation ; but his true greatness can only be estimated by a consideration of the fact that he was a fertile teacher not only of human and comparative Anatomy and Zoology but also of Physiology, and that nearly all the most distinguished German zoologists and physiologists of the period 1850 to

[1] See Agassiz, *Essay on Classification*, 1859, for an account of them.

1870 were his pupils and acknowledged his leadership.
The most striking feature about Johann Müller's
work, apart from the comprehensiveness of his point
of view, in which he added to the anatomical and
morphological ideas of Cuvier a consideration of
Physiology, Embryology, and microscopic structure,
was the extraordinary accuracy, facility, and completeness of his recorded observations.   He could do more
with a single specimen of a rare animal (*e.g.* in his
memoir on *Amphioxus*, Berlin, 1844) in the way of
making out its complete structure than the ablest of
his contemporaries or successors could do with a
plethora.   His power of rapid and exhaustive observation and of accurate pictorial reproduction was
phenomenal.   His most important memoirs, besides
that just mentioned, are those on the anatomy and
classification of Fishes, on the Cœcilians, and on the
developmental history of the Echinoderms.

A name which is apt to be forgotten in the period
between Cuvier and Darwin, because its possessor
occupied an isolated position in England and was not
borne up by any great school or university, is that of
John Vaughan Thompson, who was an army surgeon,
and when past the age of forty, being district medical
inspector at Cork (1830), took to the study of marine
Invertebrata by the aid of the microscope.   Thompson
made three great discoveries, which seem to have
fallen in his way in the most natural and simple
manner, but must be regarded really as the outcome
of extraordinary genius.   He showed that the organisms like *Flustra* are not hydroid Polyps, but of a

more complex structure resembling Molluscs, and he gave them the name "Polyzoa." He discovered the *Pentacrinus europæus*, and showed that it was the larval form of the Feather-Star *Antedon* (*Comatula*). He upset Cuvier's retention of the Cirripedes among Mollusca, and his subsequent treatment of them as an isolated class, by showing that they begin life as free-swimming Crustacea identical with the young forms of other Crustacea. Vaughan Thompson is a type of the marine zoologists, such as Dalyell, Michael Sars, P. J. Van Beneden, Claparède, and Allman, who during the present century have approached the study of the lower marine organisms in the same spirit as that in which Trembley and Schäffer in the last century, and Swammerdam in the seventeenth, gave themselves up to the study of the minute fresh-water forms of animal life.

It is impossible to enumerate or to give due consideration to all the names in the army of anatomical and embryological students of the middle third of this century whose labours bore fruit in the modification of zoological theories and in the building up of a true classification of animals. Their results are best summed up in the three schemes of classification which follow below—those of Rudolph Leuckart (*b.* 1823), Henri Milne-Edwards (1800-1884), and T. H. Huxley (*b.* 1825), all of whom individually contributed very greatly by their special discoveries and researches to the increase of exact knowledge.

Contemporaneous with these were various schemes of classification which were based, not on a consideration of the entire structure of each animal, but on the variations of a single organ, or on the really non-significant fact of the structure of the egg. All such single-fact systems have proved to be useless, and in fact departures from the true line of growth of the zoological system which was shaping itself year by year—unknown to those who so shaped it—as a genealogical tree. They were attempts to arrive at a true knowledge of the relationships of animals by "royal roads"; their followers were landed in barren wastes.

R. Leuckart's[1] classification is as follows :—

Type 1. **Cœlenterata.**

  Class I. POLYPI.
        Orders : *Anthozoa* and *Cylicozoa.*
    „ II. ACALEPHÆ.
        Orders : *Discophoræ* and *Ctenophoræ.*

Type 2. **Echinodermata.**

  Class I. PELMATOZOA.
        Orders : *Cystidea* and *Crinoidea.*
    „ II. ACTINOZOA.
        Orders : *Echinida* and *Asterida.*
    „ III. SCYTODERMATA.
        Orders : *Holothuriæ* and *Sipunculida.*

Type 3. **Vermes.**

  Class I. ANENTERÆTI.
        Orders : *Cestodes* and *Acanthocephali.*

---

[1] *Die Morphologie und die Verwandtschaftsverhältnisse der wirbellosen Thiere*, Brunswick, 1848. The *Protozoa*, recognised as a primary group by Siebold and Stannius (*Lehrbuch d. vergleich. Anatomie*, Berlin, 1845), are not included at all by Leuckart in his scheme. The name *Protozoa* was first used by Goldfuss (1809) to include microscopic animals and also the Polyps and *Medusæ*, whilst Siebold and Stannius first used it in its modern signification as comprising and limited to the *Infusoria* and *Rhizopoda.*

Class II. APODES.

Orders : *Nemertini, Turbellarii, Trematodes,* and *Hirudinei.*

„ III. CILIATI.

Orders : *Bryozoa* and *Rotifera.*

„ IV. ANNELIDES.

Orders : *Nematodes, Lumbricini,* and *Branchiati.*

Type 4. **Arthropoda.**

Class I. CRUSTACEA.

Orders : *Entomostraca* and *Malacostraca.*

„ II. INSECTA.

Orders : *Myriapoda, Arachnida* (*Acera,* Latr.), and *Hexapoda.*

Type 5. **Mollusca.**

Class I. TUNICATA.

Orders : *Ascidiæ* and *Salpæ.*

„ II. ACEPHALA.

Orders : *Lamellibranchiata* and *Brachiopoda.*

„ III. GASTEROPODA.

Orders : *Heterobranchia, Dermatobranchia, Heteropoda, Ctenobranchia, Pulmonata,* and *Cyclobranchia.*

„ IV. CEPHALOPODA.

Type 6. **Vertebrata.** (Not specially dealt with.)

The classification given by Henri Milne-Edwards [1] is as follows :—

Branch I. **Osteozoaria** or **Vertebrata.**

Sub-Branch 1. **Allantoidians.**

Class I. MAMMALIA.

Orders : (*a*) Monodelphia : *Bimana, Quadrumana, Cheiroptera, Insectivora, Rodentia, Edentata, Carnivora, Amphibia, Pachydermata, Ruminantia, Cetacea ;* (*b*) Didelphia : *Marsupialia, Monotremata.*

„ II. BIRDS.

Orders : *Rapaces, Passeres, Scansores, Gallinæ, Grallæ, Palmipedes.*

„ III. REPTILES.

Orders : *Chelonia, Sauria, Ophidia.*

---

[1] *Cours Élémentaire d'Histoire Naturelle,* Paris, 1855.

Sub-Branch 2. **Anallantoidians.**
>Class I. BATRACHIANS.
>>Orders: *Anura, Urodela, Perennibranchia, Cæciliæ.*
>
>„ II. FISHES.
>>Section 1. *Ossei.*
>>>Orders: *Acanthopterygii, Abdominales, Sub-brachii, Apodes, Lophobranchii, Plecto-gnathi.*
>>
>>Section 2. *Chondropterygii.*
>>>Orders: *Sturiones, Selachii, Cyclostomi.*

Branch II. **Entomozoa** or **Annelata.**
>Sub-Branch 1. **Arthropoda.**
>>Class I. INSECTA.
>>>Orders: *Coleoptera, Orthoptera, Neuroptera, Hymenoptera, Lepidoptera, Hemiptera, Di-ptera, Rhipiptera, Anopleura, Thysanura.*
>>
>>„ II. MYRIAPODA.
>>>Orders: *Chilognatha* and *Chilopoda.*
>>
>>„ III. ARACHNIDS.
>>>Orders: *Pulmonaria* and *Trachearia.*
>>
>>„ IV. CRUSTACEA.
>>>Section 1. *Podophthalmia.*
>>>>Orders: *Decapoda* and *Stomopoda.*
>>>
>>>Section 2. *Edriophthalmia.*
>>>>Orders: *Amphipoda, Læmodipoda,* and *Isopoda.*
>>>
>>>Section 3. *Branchiopoda.*
>>>>Orders: *Ostracoda, Phyllopoda,* and *Tri-lobitæ.*
>>>
>>>Section 4. *Entomostraca.*
>>>>Orders: *Copepoda, Cladocera, Siphonostoma, Lernæida, Cirripedia.*
>>>
>>>Section 5. *Xiphosura.*

(The orders of the classes which follow are not given in the work quoted.)

Sub-Branch 2. **Vermes.**

Class I. ANNELIDS.	Class IV. CESTOIDEA.
„ II. HELMINTHS.	„ V. ROTATORIA.
„ III. TURBELLARIA.	

Branch III. **Malacozoaria** or **Mollusca**.
    Sub-Branch 1. **Mollusca** proper.
        Class I. CEPHALOPODA.   Class III. GASTEROPODA.
          ,, II. PTEROPODA.   ,, IV. ACEPHALA.
    Sub-Branch 2. **Molluscoidea**.
        Class I. TUNICATA.   Class II. BRYOZOA.
Branch IV. **Zoophytes**.
    Sub-Branch I. **Radiaria**.
        Class I. ECHINODERMS.   Class III. CORALLARIA or
          ,, II. ACALEPHS.   POLYPI.
    Sub-Branch 2. **Sarcodaria**.
        Class I. INFUSORIA.   Class II. SPONGIARIA.

In England T. H. Huxley adopted in his lectures (1869) a classification which was in many respects similar to both of the foregoing, but embodied improvements of his own.   It is as follows :—

Sub-Kingdom I. **Protozoa**.
    Classes : RHIZOPODA, GREGARINIDA, RADIOLARIA, SPONGIDA.
Sub-Kingdom II. **Infusoria**.
Sub-Kingdom III. **Cœlenterata**.
    Classes : HYDROZOA, ACTINOZOA.
Sub-Kingdom IV. **Annuloida**.
    Classes : SCOLECIDA, ECHINODERMATA.
Sub-Kingdom V. **Annulosa**.
    Classes : CRUSTACEA, ARACHNIDA, MYRIAPODA, INSECTA, CHÆ-
        TOGNATHA, ANNELIDA.
Sub-Kingdom VI. **Molluscoida**.
    Classes : POLYZOA, BRACHIOPODA, TUNICATA.
Sub-Kingdom VII. **Mollusca**.
    Classes : LAMELLIBRANCHIATA, BRANCHIOGASTROPODA, PULMO-
        GASTROPODA, PTEROPODA, CEPHALOPODA.
Sub-Kingdom VIII. **Vertebrata**.
    Classes : PISCES, AMPHIBIA, REPTILIA, AVES, MAMMALIA.

We now arrive at the period when the doctrine of organic evolution was established by Darwin, and when naturalists, being convinced by him as they had not been by the transmutationists of fifty years' earlier date, were compelled to take an entirely new

view of the significance of all attempts at framing a
" natural " classification.

Many zoologists—prominent among them in Great
Britain being Huxley—had been repelled by the airy
fancies and assumptions of the " philosophical " morphologists. The efforts of the best minds in Zoology
had been directed for thirty years or more to ascertaining with increased accuracy and minuteness the
structure, microscopic and gross, of all accessible forms
of animals, and not only of the adult structure but of
the steps of development of that structure in the
growth of each kind of organism from the egg to
maturity. Putting aside fantastic theories, these observers endeavoured to give in their classifications a
strictly objective representation of the facts of animal
structure and of the structural relationships of animals
to one another capable of demonstration. The groups
within groups adopted for this purpose were necessarily wanting in symmetry : the whole system presented a strangely irregular character. From time to
time efforts were made by those who believed that the
Creator must have followed a symmetrical system in
his production of animals to force one or other artificial, neatly balanced scheme of classification upon the
zoological world. The last of these was that of Louis
Agassiz (*Essay on Classification*, 1859), who, whilst
surveying all previous classifications, propounded a
scheme of his own, in which, as well as in the criticisms
he applies to other systems, the leading notion is that
sub-kingdoms, classes, orders, and families have a real
existence. He held that it is possible to ascertain and

distinguish characters which are of class value, others which are only of ordinal value, and so on, so that the classes of one sub-kingdom should on paper, and in nature actually do, correspond in relative value to those of another sub-kingdom, and the orders of any one class similarly should be so taken as to be of equal value with those of another class, and have been actually so created.

The whole position was changed by the acquiescence, which became universal, in the doctrine of Darwin. That doctrine took some few years to produce its effect, but it became evident at once to those who accepted Darwinism that *the* natural classification of animals, after which collectors and anatomists, morphologists, philosophers, and embryologists had been so long striving, was nothing more nor less than a genealogical tree, with breaks and gaps of various extent in its record. The facts of the relationships of animals to one another, which had been treated as the outcome of an inscrutable law by most zoologists and glibly explained by the transcendental morphologists, were amongst the most powerful arguments in support of Darwin's theory, since they, together with all other vital phenomena, received a sufficient explanation through it. It is to be noted that, whilst the zoological system took the form of a genealogical tree, with main stem and numerous diverging branches, the actual form of that tree, its limitation to a certain number of branches corresponding to a limited number of divergences in structure, came to be regarded as the necessary consequence of the operation of the physico-chemical laws of the universe, and it was recognised that the ultimate explanation of that limitation is to be found only in the constitution of matter itself.

The first naturalist to put into practical form the consequences of the new theory, in so far as it affected zoological classification, was Ernst Haeckel of Jena (*b.* 1834) who in 1866, seven years after the publication of Darwin's *Origin of Species,* published his suggestive *Generelle Morphologie.* Haeckel introduced into classification a number of terms intended to indicate the branchings of a genealogical tree. The whole "system" or scheme of classification was termed a genealogical tree (*Stammbaum*); the main branches were termed "phyla," their branchings "sub-phyla"; the great branches of the sub-phyla were termed "cladi," and the "cladi" divided into "classes," these into sub-classes, these into legions, legions into orders, orders into sub-orders, sub-orders into tribes, tribes into families, families into genera, genera into species. Additional branchings could be indicated by similar terms where necessary. There was no attempt in Haeckel's use of these terms to make them exactly or more than approximately equal in significance; such attempts were clearly futile and unimportant where the purpose was the exhibition of lines of descent, and where no natural equality of groups was to be expected *ex hypothesi.* Haeckel's classification of 1866 was only a first attempt. In the edition of the *Natürliche Schöpfungsgeschichte* published in 1868, he made a great advance in his genealogical classification, since he now introduced the results of the extraordinary

activity in the study of Embryology which followed on
the publication of the *Origin of Species.*

The pre-Darwinian systematists since the time of
Von Baer had attached very great importance to
embryological facts, holding that the stages in an
animal's development were often more significant of
its true affinities than its adult structure. Von Baer
had gained unanimous support for his dictum, "Die
Entwickelungsgeschichte ist der wahre Lichtträger für
Untersuchungen über organische Körper." Thus J.
Müller's studies on the larval forms of Echinoderms
and the discoveries of Vaughan Thompson were appreciated. But it was only after Darwin that the cell-
theory of Schwann was extended to the Embryology
of the animal kingdom generally, and that the knowledge of the development of an animal became a
knowledge of the way in which the millions of cells of
which its body is composed take their origin by fission
from a smaller number of cells, and these ultimately
from the single egg-cell. Kölliker (*Development of
Cephalopods,* 1844), Remak (*Development of the Frog,*
1850), and others had laid the foundations of this
knowledge in isolated examples ; but it was Kowalewsky, by his accounts of the development of Ascidians
and of *Amphioxus* (1866), who really made zoologists
see that a strict and complete cellular embryology of
animals was as necessary and feasible a factor in the
comprehension of their relationships as at the beginning of the century the coarse anatomy had been
shown to be by Cuvier. Kowalewsky's work appeared between the dates of the *Generelle Morpho-*

logie and the *Schöpfungsgeschichte*. Haeckel himself, with his pupil Miklucho-Maclay, had in the meantime made studies on the growth from the egg of Sponges, —studies which resulted in the complete separation of the unicellular or equicellular Protozoa from the Sponges, hitherto confounded with them. It is this introduction of the consideration of cell-structure and cell-development which, subsequently to the establishment of Darwinism, has most profoundly modified the views of systematists, and has led in conjunction with the genealogical doctrine to the greatest activity in research,—an activity which culminated in the work (1873-1882) of F. M. Balfour, and produced the profoundest modifications in classification.

Haeckel's earlier pedigree is worth comparing with his second effort, as showing the beginning of the influence just noted. The second pedigree is as follows :—

Phyla.	Cladi.	Classes.
Protozoa.	Ovularia.	Archezoa. Gregarinæ. Infusoria.
	Blastularia.	Plancæada. Gastræada.
Zoophyta.	Spongiæ.	Porifera.
	Acalephæ.	Coralla. Hydromedusæ. Ctenophora.
Vermes.	Acœlomi.	Platyhelminthes.
	Cœlomati.	Nemathelminthes. Bryozoa. Tunicata. Rhynchocœla. Gephyræa. Rotatoria. Annelida.

Mollusca.	ACEPHALA.	Spirobranchia. Lamellibranchia.
	EUCEPHALA.	Cochlides. Cephalopoda.
Echinoderma.	COLOBRACHIA.	Asterida. Crinoida.
	LIPOBRACHIA.	Echinida. Holothuriæ.
Arthropoda.	CARIDES.	Crustacea.
	TRACHEATA.	Arachnida. Myriapoda. Insecta.
Vertebrata.	ACRANIA.	Leptocardia.
	MONORRHINA.	Cyclostoma.
	ANAMNIA.	Pisces. Dipneusta. Halisauria. Amphibia.
	AMNIOTA.	Reptilia. Aves. Mammalia.

In representing pictorially the groups of the animal
kingdom as the branches of a tree, it becomes obvious
that a distinction may be drawn, not merely between
the individual main branches, but further as to the
level at which they are given off from the main stem,
so that one branch or set of branches may be marked
off as belonging to an earlier or lower level than
another set of branches; and the same plan may be
adopted with regard to the cladi, classes, and smaller
branches. The term "grade" was introduced by
myself[1] to indicate this giving off of branches at a
higher or lower, *i.e.* a later or earlier, level of a main
stem. The mechanism for the statement of the genea-

[1] "Notes on Embryology and Classification," in *Quart. Journ. Micr.
Sci.* 1877.

THE HISTORY AND SCOPE OF ZOOLOGY

logical relationships of the groups of the animal kingdom was thus completed. Renewed study of every
group was the result of the acceptance of the genealogical idea and of the recognition of the importance
of cellular Embryology. On the one hand, the true
method of arriving at a knowledge of the genealogical
tree was recognised as lying chiefly in attacking the
problem of the genealogical relationships of the smallest
twigs of the tree, and proceeding from them to the
larger branches. Special studies of small families or
orders of animals with this object in view were taken
in hand by many zoologists. On the other hand, a
survey of the facts of cellular Embryology which were
accumulated in regard to a variety of classes within a
few years of Kowalewsky's work led to a generalisation, independently arrived at by Haeckel and myself,
to the effect that a lower grade of animals may be
distinguished, the *Protozoa* or *Plastidozoa*, which
consist either of single cells or colonies of equiformal
cells, and a higher grade, the *Metazoa* or *Enterozoa*,
in which the egg-cell by "cell division" gives rise to
two layers of cells, the endoderm and the ectoderm,
surrounding a primitive digestive chamber, the archenteron. Of these latter I proposed to distinguish two
grades,—those which remain possessed of a single
archenteric cavity and of two primary cell-layers (the
Cœlentera or *Diploblastica*), and those which by
nipping off the archenteron give rise to two cavities,
the cœlom or body-cavity and the metenteron or gut
(*Cœlomata* or *Triploblastica*). To the primitive two-
cell-layered form, the hypothetical ancestor of all

Metazoa or *Enterozoa*, Haeckel gave the name *Gastræa*; the embryonic form which represents in the individual growth from the egg this ancestral condition he called a "gastrula." The term "diblastula" has more recently been adopted in England for the "gastrula" of Haeckel. The tracing of the exact mode of development, cell by cell, of the diblastula, the cœlom, and the various tissues of examples of all classes of animals has been pursued during the last twenty years with immense activity and increasing instrumental facilities, and is still in progress.

Two names in connection with post-Darwinian taxonomy and the ideas connected with it require brief mention here. Fritz Müller, by his studies on Crustacea (*Für Darwin*, 1864), showed the way in which genealogical theory may be applied to the minute study of a limited group. He is also responsible for the formulation of an important principle, called by Haeckel " the biogenetic fundamental law," viz. that an animal in its growth from the egg to the adult condition tends to pass through a series of stages which are recapitulative of the stages through which its ancestry has passed in the historical development of the species from a primitive form ; or, more shortly, that the development of the individual (ontogeny) is an epitome of the development of the race (phylogeny). Pre-Darwinian zoologists had been aware of the class of facts thus interpreted by Fritz Müller, but the authoritative view on the subject had been that there is a parallelism between (*a*) the series of forms which occur in individual development, (*b*) the series of

existing forms from lower to higher, and (c) the series
of forms which succeed one another in the strata of
the earth's crust; whilst an explanation of this parallelism was either not attempted, or was illusively offered
in the shape of a doctrine of harmony of plan in creation.   It was the application of Fritz Müller's law of
recapitulation which gave the chief stimulus to recent
embryological investigations; and, though it is now
recognised that "recapitulation" is vastly and be-
wilderingly modified by special adaptations in every
case, yet the principle has served, and still serves, as
a guide of great value.

Another important factor in the present condition
of zoological knowledge as represented by classification
is the doctrine of degeneration propounded by Anton
Dohrn.   Lamarck believed in a single progressive
series of forms, whilst Cuvier introduced the conception
of branches.   The first post-Darwinian systematists
naturally and without reflection accepted the idea
that existing simpler forms represent stages in the
gradual progress of development,—are in fact survivors
from past ages which have retained the exact grade
of development which their ancestors had reached in
past ages.   The assumption made was that (with the
rare exception of parasites) all the change of structure
through which the successive generations of animals
have passed has been one of progressive elaboration.
It is Dohrn's merit to have pointed out[1] that this
assumption is not warranted, and that degeneration or

[1] *Ursprung der Wirbelthiere*, Leipsic, 1875 ; and Lankester, *Degeneration*, London, 1880 (reprinted in the present volume).

progressive simplification of structure may have, and
in many lines certainly has, taken place, as well as progressive elaboration and continuous maintenance of the
status quo. The introduction of this conception necessarily has had a most important effect in the attempt
to unravel the genealogical affinities of animals. It
renders the task a more complicated one ; at the same
time it removes some serious difficulties and throws a
flood of light on every group of the animal kingdom.

One result of the introduction of the new conceptions dating from Darwin has been a healthy reaction
from that attitude of mind which led to the regarding
of the classes and orders recognised by authoritative
zoologists as sacred institutions which were beyond
the criticism of ordinary men. That attitude was
due to the fact that the groupings so recognised did
not profess to be simply the result of scientific reasoning, but were necessarily regarded as the expressions
of the "insight" of some more or less gifted persons
into a plan or system which had been arbitrarily
chosen by the Creator. Consequently there was a
tinge of theological dogmatism about the whole matter.
To deny the Linnæan, or later the Cuvierian, classes
was very much like denying the Mosaic cosmogony.
At the present time systematic Zoology is entirely
free from any such prejudices, and the Linnæan taint
which is apparent even in Haeckel and Gegenbaur
may be considered as finally expunged.

A classification which expresses the probabilities
of genealogical relationships as indicated by the latest
results of investigation, is that at which every teacher

of Zoology now aims. That which at the present
moment commends itself to me is represented by the
genealogical tree and the tabular statements which
are printed at the end of this essay.[1] The chief points
in this classification are the inclusion of *Balanoglossus*
and the *Tunicata* in the phylum *Vertebrata*, the
association of the *Rotifera* and the *Chætopoda* with
the *Arthropoda* in the phylum *Appendiculata*, the
inclusion of *Limulus* and the *Eurypterina* in the
class *Arachnida*, and the total abandoning of the
indefinite and indefensible group of " *Vermes.*"

We have now traced the history of the morphography of animals so as to show that increasingly in
successive epochs independent branches of knowledge
have been brought to bear on the consideration of the
main problem, namely, the discrimination of the kinds
and the relations to one another of animal forms.
Before glancing at the history of the remaining
branches of zoological science, which have had an
independent history whilst ultimately contributory to
Taxonomy and Morphography, it may be briefly
pointed out that the accumulation of knowledge with
regard to the distribution of animal forms on the
earth's surface and in the seas has progressed simultaneously with the discrimination of the mere forms
of the species themselves, as has also the knowledge
derived from fossilised remains as to the characters
of former inhabitants of the globe. Both these subdivisions of Morphography have contributed to the

---

[1] The classification here given is not quite the same as that published in the *Encyclopædia Britannica* in 1888.

establishment of Darwinism,—the one (palæontology) by direct evidence of organic evolution in time, the other (zoo-geography) in a more indirect way.

Alfred Russell Wallace stands prominently forward as a naturalist-traveller who by his observations, chiefly on Lepidopterous Insects, in both South America and the Malay Archipelago, was led to the conclusion that a production of new species is actually going on, and that, too, by means of a process of natural selection of favourable variations. Wallace and Darwin, who each recognised cordially and fully the other's work, laid their views before the Linnean Society on the same day—1st July 1858.

The facts of the geographical distribution of animals were systematised, and great zoo-geographical provinces first clearly recognised, by P. L. Sclater [1] in 1858. The application of the Darwinian theory to the facts tabulated by Sclater, combined with a knowledge of the distribution of animals in past geological periods, has led to a full explanation of the migrations of terrestrial animals, and has furnished a striking corroboration of the sufficiency of the doctrine of organic evolution, as reformed by Darwin, to account for all the phenomena of Zoology.

The study of the marine fauna by means of the dredge and trawl had been enthusiastically prosecuted by British, French, and Scandinavian naturalists in the two decades before Darwin's book; the collection of forms, the discovery of new species, and the recording of their bathymetrical and local distribution had

[1] *Journal of the Proc. Linnæan Soc. Zool.* vol. ii. p. 130.

produced a great mass of knowledge through the
labours of E. Forbes, Gwyn-Jeffreys, Sars, Quatre-
fages, Norman, and others. The post-Darwinian
developments of this line of inquiry have been two.
In the first place, dredging and trawling have been
extended by the aid of steamships of the Norwegian,
British, American, French, and Italian navies into
greater depths than were previously supposed to contain living things. New species and genera, and a
vast extension of knowledge as to distribution, have
been the outcome of these expeditions, connected with
the names of G. O. Sars and Daniellsen in Norway, of
Alex. Agassiz in America, and of Carpenter and
Wyville Thomson in Great Britain. It is worthy of
note that the practical demand for sounding the
Atlantic in connection with the laying of the first
deep-sea telegraph-cable is what led to these explora-
tions, the first recognition of life at these great depths
in the ocean being due to Dr. Wallich, who accompanied a sounding expedition in 1860 to the North
Atlantic, and to Professor Fleeming Jenkin, who in the
same year acted as engineer in raising the submarine
cable between Sardinia and Africa, upon which living
corals were found. In the second place, the study of
marine Zoology has, since the publication of the *Origin
of Species*, been found to require more complete
arrangements in the form of laboratories and aquaria
than the isolated vacation student could bring with
him to the seaside. Seaside laboratories have come
into existence : the first was founded in France by
Coste (1859) at Concarneau (Brittany), again with a

practical end in view, viz. the study of food-fishes with an aim to pisciculture. The demand for a knowledge of the embryology of all classes of animals, and for further facts as to the structure and life-history of the minuter microscopic or very delicate forms of marine life, is what has determined the multiplication of these marine "stations." The largest and best supported pecuniarily is that founded at Naples by Anton Dohrn in 1872; others exist at Trieste, Villefranche, Cette, and at New Haven and Beaufort in the United States, whilst a large laboratory, on a scale to compare with that at Naples, has this year (1888) been opened at Plymouth by the Marine Biological Association of the United Kingdom.

Another result of the stimulus given to zoological research by Darwin's work is the undertaking of voyages to distant lands by skilled anatomists for the purpose of studying on the spot, and with all the advantages of abundant and living material, the structure, and especially the embryology, of rare and exceptionally interesting forms of animal life. In the pre-Darwinian period of this century zoologists who were convinced of the importance of anatomical and embryological study were still content to study specimens immersed in spirit and brought home, often imperfectly preserved, by unskilled collectors, or to confine their attention to such species as could be procured in Europe. Before Cuvier, as we have already pointed out, attention was, with rare exceptions, limited to the dried skeletons and external forms of animals. Now, however, the enterprising

zoologist goes to the native land of an interesting
animal, there to study it as fully as possible. The
most important of these voyages has been that of W.
H. Caldwell of Cambridge to Australia (1885-1886) for
the purpose of studying the embryology of the *Monotrema* and of *Ceratodus*, the fish-like *Dipnoon*, which
has resulted in the discovery that the *Monotrema* are
oviparous. Similarly Adam Sedgwick proceeded to
the Cape in order to study *Peripatus*, Bateson to the
coast of Maryland to study *Balanoglossus*, and the
brothers Sarassin to Ceylon to investigate the embryology of the *Cæcilia*.

The task of the zoologist has changed and developed in every succeeding period. Pure morphography
has long ceased to be a chief line of research; and
now even the preoccupation produced by the addition
to it of the study of cellular Embryology is about to
undergo a modification by the demand for knowledge
of the facts of heredity and adaptation in greatly
extended detail.

Zoo-Mechanics, Zoo-Physics, Zoo-Chemistry

The development of that knowledge of the structure of the human body, and of the chemical and
physical processes going on in it, which is necessary
for the purposes of the medical art, forms a distinct
history, which has both influenced and been influenced
by that of other branches of Zoology. The study of
the structure and composition of the body of man and
of the animals nearest to him was until fifty years ago

one with the inquiry into the activities of those parts, and indeed the separation of Anatomy and Physiology has never been really carried out. For convenience of teaching, the description of the coarser anatomy of the human body has been in modern universities placed in the hands of a special professor, theoretically condemned to occupy himself with the mere formal details of structure, whilst the professor of Physiology has usually retained what is called "microscopic anatomy," and necessarily occupies himself with as much structural Anatomy as is required for a due description of the functions of organs and the properties of tissues. It would seem that in our medical schools and universities these arrangements should be reconsidered. Anatomy and Physiology should be re-united and subdivided as follows,—(1) Physiology with Anatomy in relation to Physiology, (2) Anatomy in relation to surgery and medical diagnosis,—the former being a science, the latter a piece of technical training in rule of thumb.

Physiological Anatomy or anatomical Physiology has its beginnings in Aristotle and other observers of antiquity. The later Græco-Roman and the Arabian physicians carried on the traditional knowledge and added to it. Galen dominated the Middle Ages. The modern development begins with Harvey and with the Italian school in which he studied. Its great names are Fabricius of Acquapendente (1537-1619), Vesalius (1514-1564), Eustachius (c. 1500-1574), Riolan (1577-1657), Severino (1580-1656). The history of the discovery of the circulation of the blood

and of the controversies connected with it gives
an interesting and sufficient presentation of the
anatomico-physiological knowledge of the period.
The foundation of the scientific academies and the
records of their publications furnish thenceforward
a picture of the progress in this study.   As an
early anatomist Willis (1621-1675), professor of
physic in Oxford, deserves notice for his work on the
anatomy of the human brain, the plates for which
were drawn by young Christopher Wren, the prodigy
of Oxford common-rooms, who later built St. Paul's
Cathedral.   The Royal Society, in its early days when
Wren was a fellow, met at Gresham College whenever
the professor of physic there could obtain a human
body for dissection, and amongst its earliest records
are the memoirs of Tyson on the anatomy of the
Chimpanzee and the experiments on transfusion of
blood, extirpation of the spleen, and such like inquiries.

Marcello Malpighi (1628-1694) and Anton van
Leeuwenhoek (1632-1723) were the first to introduce
the microscope into anatomical research.   Malpighi
first used the injection of blood-vessels on a large scale,
and moreover is to be credited with having first conceived that there is a definite relation of the structure
of lower kinds of animals to that of higher and more
elaborate kinds, and that this relation is one of gradual
transition, so that lower animals are not to be regarded
as isolated and arbitrary existences, but are really
simpler exhibitions of the same kind of structure and
mechanism which occurs in higher animals.   It is this
conception which later developed into the theory of

an actual transmutative development of lower into
higher organisms. Leeuwenhoek discovered the red
blood corpuscles of Vertebrates, saw the circulation in
the capillaries of the Frog's foot, described the fibrillar
structure and cross-striping of muscular fibre, the
tubular structure of dentine, the scales of the epi-
dermis, the fibres of the lens, and the spermatozoa,
these last having been independently discovered at
Leyden in 1677 by Ludwig Ham of Stettin. The
spermatozoa were regarded by the "animalculists" as
the fully formed but minute young which had to be
received in the egg, in order to be nourished and
increase in size, and were hailed as a decisive blow to
Harvey's doctrine of epigenesis and his dictum "omne
vivum ex ovo." Albrecht von Haller was the champion
of the so-called "evolutionists" in the eighteenth century, better called "præformationists." Haller wrote,
"There is no such thing as development! No part
of the animal body is made before another; all are
simultaneously created." A corollary of this doctrine
was that the germ contains the germs of the next
generation, and these of the next, and so *ad infinitum.*
It was calculated that Eve at her creation thus contained within her 200,000 millions of human germs.
This was the view of the "ovists," who regarded the
egg as the true germ, whilst the "animalculists," who
regarded the spermatozoon as the essential germ, would
have substituted Adam for Eve in the above calculation. These fanciful conceptions—containing as they
do a share of important truth—were opposed by
Caspar Friedrich Wolff, who in his doctorate disserta-

tion (1759) maintained that the germ is a structureless particle, and acquires its structure by "epigenesis" or gradual development.   Wolff has proved to be nearer the truth than Haller; but modern conceptions as to the molecular structure of the egg-protoplasm point to a complexity as great as that imagined by the evolutionists.   Later it was maintained that the spermatozoa are parasitic animalcules, and this view prevailed for 150 years, so that in the *Physiology* of Johann Müller (1842) we read, "Whether the spermatozoa are parasitic animalcules or living parts of the animal in which they occur cannot at present be stated with certainty."

Physiology in the eighteenth century could only proceed by means of inferences from purely anatomical observation, aided by imaginative conceptions which had no real basis.   The explanation of the processes of life in the animal body was waiting for that progress in the knowledge of physics and chemistry which at last arrived, and gave a new impulse to investigation. Albrecht von Haller (1708-1777) was the first to apply experimental methods to the determination of the functions of the various organs made known by anatomists, and from him we may trace a bifurcation in the tendencies of medical men who occupied themselves with the study of the structure and functions of the animal organism.   The one class proceeded more and more in the direction of comparative anatomy, the other in the direction of exact analysis and measurement of both the structure and properties of the organs of Vertebrate animals allied to man and of man himself.

John Hunter (1728-1793) is the most striking figure of this epoch in the relation of medicine to general zoological progress; his museum, preserved in Lincoln's Inn Fields, London, by the combined action of the state and the Royal College of Surgeons, is an abiding record of the historical progress of biological science. Hunter collected, dissected, and described not only higher but lower animals, with the view of arriving at a knowledge of the function of organs by the most extensive and systematic survey of their modifications in all kinds of animals. His purpose was that of the physiologist and medical man, but he made great contributions to the general knowledge of animal structure. The same class of investigations, when taken up by Cuvier from the point of view of systematic Zoology and Morphology, led to a reconstruction of classification and laid the foundation of anatomical Zoology. Hunter was the younger brother of William Hunter, who also formed an important museum, still preserved in Glasgow. Hunter classified the organs of animals into those which subserve the preservation of the individual, those which subserve the preservation of the species, and those which are the means of relation with the outer world, and he arranged his museum of dissections and preparations on this plan.

The great progress of chemistry at the end of the eighteenth and the beginning of the nineteenth century was followed by an application of chemical laws and chemical methods to the study of animal life. Curiously enough, as showing how deeply interwoven are the various lines of scientific progress, Priestley in his discovery of oxygen was as much concerned in the study of a chlorophyll-bearing Protozoon, *Euglena viridis*, as in that of the red oxide of mercury; and the interest in "vital spirits" as a physiological factor was an important stimulus to those researches which produced modern chemical knowledge.

The purely anatomical side of physiological progress is marked in the beginning of the nineteenth century by the work of Bichat (1771-1802), who distinguished by naked-eye characters the different structural materials of which the organs of man and the higher animals are built, and thus founded in first outline the science of Histology. By the end of the first quarter of this century it had become clear to the minds of the anatomico-physiological students of animal life that the animal body was subject to the same physical laws as other matter, although it was still held that some additional and mysterious agent —so-called "vitality"—was at work in living bodies. It had become clear that animal material could be investigated chemically, and that the processes of digestion, assimilation, respiration, and secretion were chemical processes.

To a considerable extent the chemical composition and properties of the tissues, and the chemical nature of the various changes of life and of putrefaction after death, had been investigated, but one step was yet to be taken which brings the study of ultimate structure, chemical activity, form, and the development of form to a single focus. This was taken by Theodore Schwann (1810-1881), who in 1839 published his epoch-making cell-theory. Schwann was a pupil of Johann Müller, and there can be little doubt that the ideas of the pupil are to be credited in some measure to the master. Schwann took up the thread of microscopic investigation which had been sedulously pursued by a distinct line of students since the days of Hook and Leeuwenhoek, and had resulted in a general doctrine among botanists of the cellular structure of all the parts of plants. Schwann showed not only that plants are uniformly built up by these corpuscular units (of which Robert Browne in 1833 had described the peculiar nucleated structure), but that all animal tissues are also so built up. That, however, was not Schwann's chief point. The cell-theory for which he is famous is this, that the substance of the individual cell is the seat of those chemical processes which seen *en masse* we call life, and that the differences in the properties of the different tissues and organs of animals and plants depend on a difference in the chemical and physical activity of the constituent cells, resulting in a difference in the form of the cells and in a concomitant difference of function. Schwann thus pointed to the microscopic cell-unit as the thing to be studied in order to arrive at a true knowledge of the processes of life and the significance of form. In founding the study of cell-substance (or protoplasm, as it was subsequently called by Max Schultze in 1861, adopting the name used by botanists for vegetable cell-contents) Schwann united two lines of inquiry, viz. that of minute investigation of structure and development and that of zoo-chemistry and zoo-physics. He spent a large part of the next forty years in an attempt to penetrate further into the structure of cell-substance ; he hoped to be able to find in cell-substance ultimate visible molecules, a knowledge of the arrangement and characters of which would explain the varying properties of protoplasm.

It is not a little remarkable that Schwann, who thus brought about the union of physiological and morphological study by his conception of cell-substance, should also have been the initiator of that special kind of experimental investigation of the physical properties of tissues by the exact methods used by physicists which, by the aid of the kymographion, the thermo-electric pile, and the galvanometer, has been so largely pursued during the last thirty years in our physiological laboratories. It is perhaps less surprising that Schwann, who had so vivid a conception of the activity and potentialities of the cell-unit, should have been the discoverer of the immensely important fact that putrefaction and fermentation are not the consequences of death but of life, and that without the presence of living *Bacteria* putrefaction does not occur, whilst he also is the discoverer of the fact that the yeast which causes alcoholic fermentation is a mass of unicellular living organisms.

From Schwann's time onward the cell became more and more the point of observation and experiment in the progress of both Morphography and Physiology. It was soon shown, chiefly through Kölliker

and Remak, that all cells originate by fission from
pre-existing cells,—a fact unknown to Schwann,—
and the doctrine " omnis cellula e cellula " was established.   It was also demonstrated that the Mammalian
egg discovered by Von Baer was a typical nucleated
cell, and that all animals, and plants also (this generalisation took thirty years to establish), take their
origin from an egg, which is in essence and in fact a
single nucleated cell.   The doctrine of Harvey, " omne
vivum ex ovo," thus received its most ample justifica-
tion.   The study of " growth from the egg " became
necessarily a study of the multiplication by fission of
the egg-cell and its fission-products, their arrangement
in layers, and the chemical metamorphosis of their
substance and exudations.   This study, as well as
the allied investigation of the cell-structure of the
adult tissues, was immensely facilitated by methods
of hardening, staining, section-cutting, and clarifying
which grew up after Schwann's time, and have their
present highest development in the automatic micro-
tome of Caldwell, which can be worked by a motor,
and delivers consecutive sections of animal tissues or
embryos $\frac{1}{4000}$th of an inch thick, arranged in the
form of ribbons, ready for examination with the microscope, at the rate of one hundred or more per minute.
Stricker of Vienna was the first to embed embryos in
waxy material for the purpose of cutting thin sections
of them, about twenty-five years ago, and R. Leuckart
of Leipsic was subsequently the first to employ this
method in the study of the structure of small Invertebrata.

The knowledge of the anatomical facts of cellular development and cellular structure necessarily gave immensely increased precision to the notion of gradation of structure in the animal series from simple to complex, and rendered Darwin's doctrine the more readily accepted. It was not, however, until after Darwin's date (1859) that the existence of unicellular animals was fully admitted, and the general facts of cellular Embryology established throughout the animal kingdom.

Similarly cellular Physiology, by establishing the conception of a simple optically homogeneous cell-substance as the seat of the activities which we call "life," rendered it possible to accept the suggestion of a simple "substance of life" which might have been evolved from simpler non-living matter by natural processes depending on physical and chemical laws. It is noteworthy that Darwin himself appears not to have been influenced directly by any such physiological or chemico-physical doctrine as to "protoplasm" or cell-substance. Nevertheless the way was prepared for the reception of Darwin's theory by this state of physiological knowledge.

The word "protoplasm" requires a little further notice. Protoplasm was applied by Von Mohl and by Max Schultze to the slimy substance of the cell, including therein both the general thinner material and the nucleus. It is, as Roscoe remarked at Manchester (*Brit. Ass. Address*, 1887), a structure and not a chemical body. Nevertheless gradually physiologists have come to use the word "protoplasm" for *one* of the chemical substances of which Schultze's protoplasm is a structural mixture—namely, that highest point in the chemical elaboration of the molecule which is attained within the protoplasm, and up to which some of the chemical bodies present are tending, whilst others are degradation products resulting from a downward metamorphosis of portions of it. This intangible, unstable, all-pervading element of the protoplasm cannot at present be identified with any visibly separable part of the cell-substance, which consists of a hyaline denser network of excessive tenuity, of a less dense hyaline liquid, and of finest and less fine granules of varying chemical nature. This "critical" substance, sometimes called "true protoplasm," should assuredly be recognised by a distinct name "plasmogen," whilst protoplasm retains its structural connotation.

The study of the process of fertilisation and of the significance in that process of the distinct parts of the sperm-cell and egg-cell—the separate fibrillæ and granules of the nuclei of those cells—at the present moment forms one of the engrossing subjects of zoological investigation.[1] Not less important is the descent, as it were, of physiological investigation in relation to every organ into the arena of the cell : digestion, secretion, muscular contraction, nerve action, all are now questions of Plasmology, or the study of cell-substance founded by Schwann.

[1] See the memoirs of Weismann on Heredity and F. M. Balfour's *Embryology.*

General Tendency of Zoology since Darwin

The serious and broadly-based study of bionomics which was introduced by Darwin, and in his hands gave rise to the doctrine of natural selection, by which the hypothesis of the origin of species by gradual transmutation in the natural process of descent from ancestral forms was established as a scientific doctrine, can hardly be said to have had any history. Buffon (1707-1788) alone among the greater writers of the three past centuries emphasised that view of living things which we call "bionomics." Buffon deliberately opposed himself to the mere exposition of the structural resemblances and differences of animals, and, disregarding classification, devoted his treatise on natural history to a consideration of the habits of animals and their adaptations to their surroundings, whilst a special volume was devoted by him to the subject of reproduction. In special memoirs on this or that animal, and in a subordinate way in systematic works, material is to be found helping to build up a knowledge of bionomics, but Buffon is the only prominent writer who can be accorded historic rank in this study.[1] The special study of man in these relations—such as is concerned with the statistics of population—must be considered as having contributed very importantly to Darwin's wider study of bionomics in general. The work of Malthus *On Popula-*

---

[1] The main literary sources made use of by Darwin are the magazines and treatises of horticulturists, farmers, pigeon-fanciers, and the like, in fact what is comprised in the *Field* newspaper.

tion (1798) exercised the most important influence on Darwin's thought, as he himself tells us, and led him to give attention to the facts of animal population, and so to discover the great moving cause of natural selection—the struggle for existence. Darwin may be said to have founded the science of bionomics, and at the same time to have given new stimulus and new direction to Morphography, Physiology, and Plasmology, by uniting them as contributories to one common biological doctrine—the doctrine of organic evolution—itself but a part of the wider doctrine of universal evolution based on the laws of physics and chemistry.

The full influence of Darwin's work upon the progress and direction of zoological study has not yet been seen. The immediate result has been, as pointed out above, a reconstruction of the classification of animals upon a genealogical basis, and an investigation of the individual development of animals in relation to the steps of their gradual building up by cell-division, with a view to obtaining evidence of their genetic relationships. On the other hand, the studies which occupied Darwin himself so largely subsequently to the publication of the *Origin of Species*, viz. the explanation of animal (and vegetable) mechanism, colouring, habits etc., as advantageous to the species or to its ancestors—in fact, the new Teleology,—has not yet been so vigorously pursued as it must be hereafter. The most important work in this direction has been done by Fritz Müller (*Für Darwin*), by Herman Müller (*Fertilisation of Plants by Insects*), and by August Weismann (memoirs translated by Meldola). Here and there observations are from time to time published, but no large progress has yet been made, probably on account of the fact that animals are exceedingly difficult to keep under observation, and that there is no provision in universities and like institutions for the pursuit of these inquiries, or even for their academic representation. More has been done with plants than with animals in this way since Darwin, probably owing to the same cause which has, ever since the revival of learning, given Botany a real advantage over Zoology, namely, the existence of "physick" gardens, now become "botanical" gardens, and the greater ease of management, experiment, and observation in the case of plants than in that of animals. It is true that zoological gardens have existed for the last fifty years in all large European cities, but these have always been conducted with a view to popular exhibition ; and, even where scientific influences have been brought to bear on their management, they have been those of the morphographer and systematist rather than of the bionomist. Moreover, zoological gardens have never been part of the equipment of the university professor of Zoology, as it may be hoped in future will be the case. The foundation of marine biological laboratories under the control of scientific zoologists offers a prospect of true bionomic observation and experiment on an increased scale in the near future, and, were such laboratories founded in our universities and provided with the necessary appliances for keeping terrestrial and freshwater animals, as well as marine forms, alive and under observation in conditions resembling as nearly as possible those of nature, a step would have been taken towards carrying on the study of Bionomics which cannot long be delayed. It seems to be even more important that the academic curriculum of Zoology should not, by mere mechanical adhesion to the old lines of Morphography, and experimental research on the chemical and physical properties of tissues and organs, confine the attention and training of young students to what are now, comparatively speaking, the less productive lines of research.

If we turn to the other branch of Bionomics, that concerned with the laws of variation and heredity (Thremmatology), we find that since Darwin, and independently of his own work, there has been a more obvious progress than in Teleology. In the first place, the continued study of human population has thrown additional light on some of the questions involved, whilst the progress of microscopical research in the hands of Bütschli, Hertwig, Balfour, and August Weismann promises to give us a clear foundation as to the structural facts connected with the origin of the egg-cell and sperm-cell and the process of fertilisation. This is not the fitting place in which to give a sketch of the doctrines and hypotheses of Thremmatology. They may be gathered from Darwin's writings, more especially the *Origin of Species* and *Animals and Plants under Domestication*.[1] They relate to

---

[1] The reader is also referred to Ribot's *L'Hérédité*, and the writings of Charles Darwin's cousin, Francis Galton.

the causes of variation in animals and plants, the
laws of the transmission of parental characters, the
share of each parent in the production of the characters of the offspring, atavism, and the relations of
young to parents as to number, sex, nourishment, and
protection.

An important development of Darwin's conclusions is actually in progress and deserves special
notice here, as it is the most distinct advance in the
department of Bionomics since Darwin's own writings,
and at the same time touches questions of fundamental interest. The matter strictly relates to the
consideration of the " causes of variation," and is as
follows.

The fact of variation is a familiar one. No two
animals, even of the same brood, are alike : whilst
exhibiting a close similarity to their parents, they yet
present differences, sometimes very marked differences,
from their parents and from one another. Lamarck
had put forward the hypothesis that structural alterations acquired by a parent in the course of its life are
transmitted to the offspring, and that, as these structural alterations are acquired by an animal or plant
in consequence of the direct action of the environment, the offspring inheriting them would as a consequence not unfrequently start with a greater fitness
for those conditions than its parents started with. In
its turn, being operated upon by the conditions of
life, it would acquire a greater development of the
same modification, which it would in turn transmit to
its offspring. In the course of several generations,

Lamarck argued, a structural alteration amounting to such difference as we call "specific" might be thus acquired. The familiar illustration of Lamarck's hypothesis is that of the giraffe, whose long neck might, he suggested, have been acquired by the efforts of a primitively short-necked race of herbivores, who stretched their necks to reach the foliage of trees in a land where grass was deficient, the effort producing a distinct elongation in the neck of each generation, which was then transmitted to the next. This process is known as "direct adaptation"; and there is no doubt that such structural adaptations are acquired by an animal in the course of its life. Whether such acquired characters can be transmitted to the next generation is a separate question. It was not proved by Lamarck that they can be, and, indeed, never has been proved by actual observation. Nevertheless it has been assumed, and also indirectly argued, that such acquired characters *must* be transmitted. Darwin's great merit was that he excluded from his theory of development any *necessary* assumption of the transmission of acquired characters. He pointed to the admitted fact of congenital variation, and he showed that these variations to all intents and purposes have nothing to do with any characters acquired by the parents, but are arbitrary and, so to speak, non-significant. Their causes are extremely difficult to trace in detail, but it appears that they are largely due to a "shaking up" of the living matter which constitutes the fertilised germ or embryo-cell, by the process of mixture in it of the substance of two cells,—

the germ-cell and the sperm-cell,—derived from two different individuals.   Other mechanical disturbances may assist in this production of congenital variation. Whatever its causes, Darwin showed that it is all-important.   In some cases a pair of animals produce ten million offspring, and in such a number a large range of congenital variation is possible.   Since on the average only two of the young survive in the struggle for existence to take the place of their two parents, there is a selection out of the ten million young, none of which are exactly alike, and the selection is determined in nature by the survival of the congenital variety which is fittest to the conditions of life.   Hence there is no *necessity* for an assumption of the perpetuation of direct adaptations.   The selection of the fortuitously (fortuitously, that is to say, so far as the conditions of survival are concerned) produced varieties is sufficient, since it is ascertained that they will tend to transmit those characters with which they themselves were born, although it is *not* ascertained that they could transmit characters *acquired* on the way through life.   A simple illustration of the difference is this : a man born with four fingers only on his right hand is ascertained to be likely to transmit this peculiarity to some at least of his offspring ; on the other hand, there is not the slightest ground for supposing that a man who has had one finger chopped off, or has even lost his arm at any period of his life, will produce offspring who are defective in the slightest degree in regard to fingers, hand, or arm.   Darwin himself, apparently influenced not merely by the

consideration of certain classes of facts which seem
to favour the Lamarckian hypothesis, but also by a
respect for the general prejudice in its favour and for
Mr. Herbert Spencer's authority, was of the opinion
that acquired characters are *in some cases* transmitted.
It should be observed, however, that Darwin did not
attribute an essential part to this Lamarckian hypothesis of the transmission of acquired characters,
but expressly assigned to it an entirely subordinate
importance.

The new attitude which has been taken since
Darwin on this question is to ask for evidence of this
asserted transmission of acquired characters. It is
held[1] that the Darwinian doctrine of selection of
fortuitous congenital variations is sufficient to account
for all cases, that the Lamarckian hypothesis of transmission of acquired characters is not supported by
experimental evidence, and that the latter should
therefore be dismissed. Weismann has also ingeni-
ously argued from the structure of the egg-cell and
sperm-cell, and from the way in which, and the period
at which, they are derived in the course of the growth
of the embryo from the egg—from the fertilised
egg-cell—that it is impossible (it would be better
to say highly improbable) that an alteration in
parental structure *could* produce any exactly representative change in the substance of the germ or
sperm-cells.

It does not seem improbable that the doctrine of
organic evolution will thus become pure Darwinism

[1] Weismann, *Vererbung*, etc. 1886.

and be entirely dissociated from the Lamarckian heresy.

The one fact which the Lamarckians can produce in their favour is the account of experiments by Brown-Séquard, in which he produced epilepsy in guinea-pigs by section of the large nerves or spinal cord, and in the course of which he was led to believe that in a few rare instances the artificially produced epilepsy was transmitted. This instance does not stand the test of criticism. It is not clear whether the guinea-pigs operated upon had or had not already a constitutional tendency to epilepsy, and it is not clear in what proportion of cases the supposed transmission took place, and whether any other disease accompanied it. On the other hand, the vast number of experiments in the cropping of the tails and ears of domestic animals, as well as of similar operations on man, are attended with negative results. No case of the transmission of the results of an injury can be produced. Stories of tailless kittens, puppies, and calves, born from parents one of whom had been thus injured, are abundant, but they have hitherto entirely failed to stand before examination.

Experimental researches on this question are most urgently needed, but they are not provided for either in the morphographical or physiological laboratories of our universities.

Whilst simple evidence of the fact of the transmission of an acquired character is wanting, the *a priori* arguments in its favour break down one after another when discussed. The very cases which are

advanced as only to be explained on the Lamarckian
assumption are found on examination and experiment
to be better explained, or only to be explained, by the
Darwinian principle. Thus the occurrence of blind
animals in caves and in the deep sea was a fact which
Darwin himself regarded as best explained by the
atrophy of the organ of vision in successive generations through the absence of light and consequent
disuse, and the transmission (as Lamarck would have
supposed) of a more and more weakened and structu-
rally impaired eye to the offspring in successive
generations, until the eye finally disappeared. But
this instance can, I think, be fully explained by the
theory of natural selection acting on congenital fortuitous variations. Many animals are thus born with
distorted or defective eyes whose parents have not
had their eyes submitted to any peculiar conditions.
Supposing a number of some species of Arthropod or
Fish to be swept into a cavern or to be carried from
less to greater depths in the sea,—those individuals
with perfect eyes would follow the glimmer of light
and eventually escape to the outer air or the shallower
depths, leaving behind those with imperfect eyes to
breed in the dark place. A natural selection would
thus be effected. In every succeeding generation this
would be the case, and even those with weak but still
seeing eyes would in the course of time escape, until
only a pure race of eyeless or blind animals would be
left in the cavern or deep sea. Experiments and
inquiries with regard to transmission of acquired
characters are in progress; amongst those who have

occupied themselves with this subject are August
Weismann of Freiburg and E. B. Poulton of Oxford.

It has been argued that the elaborate structural
adaptations of the nervous system which are the
corporeal correlatives of complicated instincts must
have been slowly built up by the transmission to
offspring of acquired experience, that is to say, of
acquired brain structure. At first sight it appears
difficult to understand how the complicated series of
actions which are definitely exhibited as so-called
" instincts " by a variety of animals can have been due
to the selection of congenital variations, or can be
otherwise explained than by the transmission of habits
acquired by the parent as the result of experience,
and continuously elaborated and added to in successive generations. It is, however, to be noted, in the
first place, that the imitation of the parent by the
young possibly accounts for some part of these complicated actions, and, secondly, that there are cases in
which curiously elaborate actions are performed by
animals as a characteristic of the species, and as sub-
serving the general advantage of the race or species,
which, nevertheless, can *not* be explained as resulting
from the transmission of acquired experience, and
must be supposed to be due to the natural selection of
a fortuitously developed habit which, like fortuitous
colour or form variation, happens to prove beneficial.
Mr. Poulton has insisted upon the habits of " sham-
ming dead " and the combined posturing and colour
peculiarities of certain caterpillars (Lepidopterous
larvæ) which cause them to resemble dead twigs or

similar surrounding objects. The advantage to the animal of this imitation of surrounding objects is that it escapes the pursuit of (say) a bird which would, were it not deceived by the resemblance, attack and eat the caterpillar. Now it is clear that preceding generations of caterpillars cannot have acquired this habit of posturing by experience. Either the caterpillar postures and escapes, or it does not posture and is eaten ; it is not half eaten and allowed to profit by experience. We seem to be justified in assuming that there are many movements of stretching and posturing possible to caterpillars, and that some caterpillars had a congenital fortuitous tendency to one position, some to another, and, finally, that among all the variety of habitual movements thus exhibited one has been selected and perpetuated because it coincided with the necessary conditions of safety, since it happened to give the caterpillar an increased resemblance to a twig.

The view that instinct is the hereditarily fixed result of habit derived from experience has hitherto dominated all inquiry into the subject, but we may now expect to see a renewed and careful study of animal instincts carried out with the view of testing the applicability to each instance of the pure Darwinian theory without the aid of Lamarckism.

The whole of this inquiry has special importance in regard to mankind, since the great questions of influence of race and family as opposed to the influence of education are at issue. If pure Darwinism is to be accepted, then education has no value in directly affecting the mental or physical features of the race, but only in affecting those of the individual. Were acquired characters really and fully transmitted, then every child born would inherit the knowledge of both its parents more or less completely, and from birth onwards would be able to add to its inherited stock, so that the progress of the race in mental acquirements would be prodigiously more rapid than it is. On the other hand, if we exclude the Lamarckian hypothesis, peculiarities of mind and body congenitally established in a race or a family acquire increased significance, for they cannot be got rid of by training, but are bound to reappear if the stock which exhibits them is allowed to breed. It seems that the laws of Thremmatology may eventually give to mankind the most precise directions, not only as to how to improve the breeds of plants and animals, but as to how to improve the human stock. It is not a little remarkable that the latest development of zoological science should favour that respect to breeding which is becoming less general than it was, and should tend to modify the current estimate of the results of popular education.

The relation of Darwinism to general philosophy and of the history of Zoology to philosophical doctrines would form one of the most interesting chapters which might be written on the subject of this article. It belongs, however, rather to the history of philosophy than to that of Zoology. Undoubtedly the conceptions of mankind at different periods of history with regard to cosmogony, and the relations of God, Nature, and Man, have had a very marked influence upon the

study of Zoology, just as in its turn the study of
Zoology has reacted upon those conceptions.

In this, as in other phases of mental development,
the ancient Greeks stand out in the most striking
manner as possessing what is sometimes called the
modern spirit. The doctrine of evolution is formulated in unmistakable terms by Heraclitus and other
philosophers of antiquity. Not only so, but the direct
examination of nature, including the various forms of
animal life, was practised by Aristotle and his disciples in a spirit which, though not altogether free
from prejudice, was yet far more like that which
actuated the founders of the Royal Society less than
three hundred years ago than anything which was
manifested in the two thousand years intervening
between that date and the time of Alexander the
Great. The study of Zoology in the Middle Ages
was simply a fantastic commentary on Aristotle and
the records of animals in the various books of the
Bible, elaborated as part of a peculiar system of
mystic philosophy, which has more analogy with the
fetichism and totem worship of savage races than with
any Greek or modern conceptions. So far as philosophy affected the study of Zoology in the beginning
of the modern period, its influence was felt in the
general acceptance of what has been called the
Miltonic cosmogony,—namely, that interpretation of
the Mosaic writings which is set forth by the poet
Milton, and of which the characteristic is the conception of the creation of existing things, including living
things, nearly or just as they are, by a rapid succession

of "fiats" delivered by an anthropomorphic Creator. It was not until the end of the eighteenth century that Schelling (as quoted above) conceived that unity of nature and general law of development which is now called the doctrine of evolution.

In England Erasmus Darwin (*Zoonomia*, published in 1794-1796), in France Lamarck (*Philosophie Zoologique*, 1809) and Geoffroy Saint-Hilaire (*Principes de Philosophie Zoologique*, 1830), and in Germany Oken (*Lehrbuch der Natur-Philosophie*, 1809-1811), Goethe (*Zur Natur Wissensch.* Stuttgart, 1817), and Treviranus (*Biologie*, 1802-1805) were the authors of more or less complete systems of a philosophy of nature in which living things were regarded as the outcome of natural law, that is, of the same general processes which had produced the inanimate universe. The "Natur-philosophen," as they were called in Germany, demand the fullest recognition and esteem. But, just in proportion as the "Natur-philosophen" failed to produce an immediate effect on the study of Zoology by their theory of natural development, so was the doctrine of evolution itself deprived of completeness and of the most important demonstration of its laws by the long-continued delay in the final introduction of Biology into the area of that doctrine.

Darwin by his discovery of the mechanical principle of organic evolution, namely, the survival of the fittest in the struggle for existence, completed the doctrine of evolution, and gave it that unity and authority which was necessary in order that it should reform the whole range of philosophy. The detailed consequences of that new departure in philosophy have yet to be worked out.   Its most important initial conception is the derivation of man by natural processes from ape-like ancestors, and the consequent derivation of his mental and moral qualities by the operation of the struggle for existence and natural selection from the mental and moral qualities of animals.   Not the least important of the studies thus initiated is that of the evolution of philosophy itself. Zoology thus finally arrives through Darwin at its crowning development : it touches and may even be said to comprise the history of man, Sociology, and Psychology.[1]

[1] The chief sources of information on the subject of the foregoing article are the following : Engelmann, *Bibliotheca Historico-Naturalis*, vol. i. 1846 (being a list of the separate works and academical memoirs relating to Zoology published between 1700 and 1846) ; Carus and Engelmann, *Bibl. Zoologica*, Leipsic, 1861 (a similar list of memoirs in periodicals published between 1846 and 1861) ; J. V. Carus, *Gesch. d. Zoologie*, Munich, 1872 ; and L. Agassiz, *An Essay on Classification*, London, 1859.

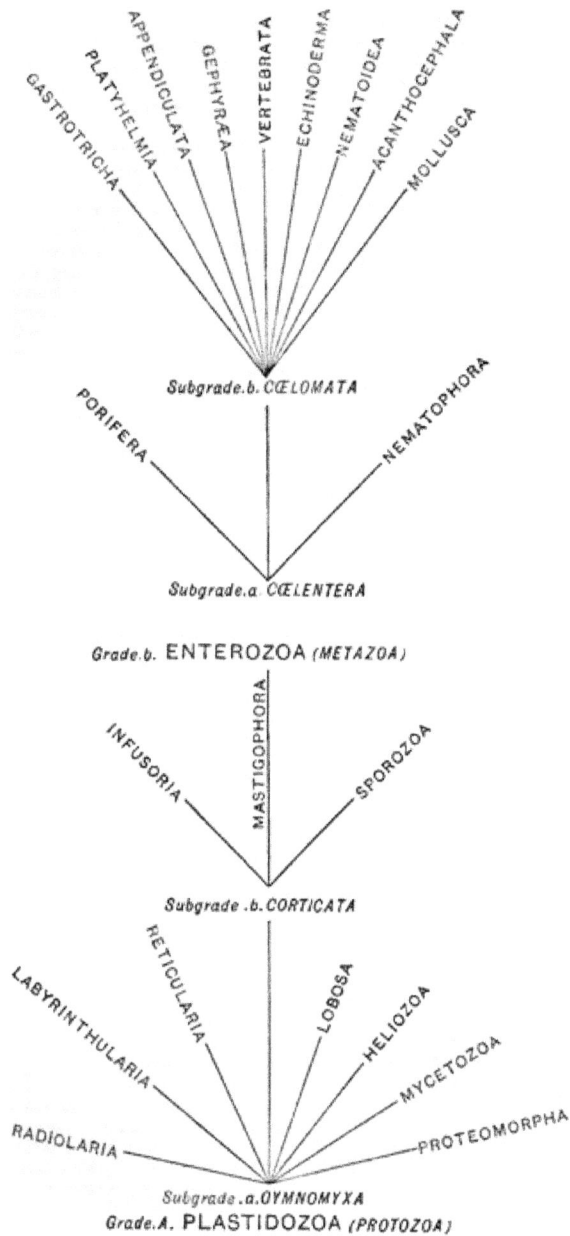

GASTROTRICHA
PLATYHELMIA
APPENDICULATA
GEPHYRÆA
VERTEBRATA
ECHINODERMA
NEMATOIDEA
ACANTHOCEPHALA
MOLLUSCA

Subgrade.b. CŒLOMATA

PORIFERA

NEMATOPHORA

Subgrade.a. CŒLENTERA

Grade.b. ENTEROZOA *(METAZOA)*

INFUSORIA
MASTIGOPHORA
SPOROZOA

Subgrade .b. CORTICATA

LABYRINTHULARIA
RETICULARIA
LOBOSA
HELIOZOA
MYCETOZOA
RADIOLARIA
PROTEOMORPHA

Subgrade .a. OYMNOMYXA
Grade.A. PLASTIDOZOA *(PROTOZOA)*

Phyla and Classes of the *PLASTIDOZOA*
(syn. *PROTOZOA*).

Sub-Grade *a.*—**Gymnomyxa.**

Class.

Phylum PROTEOMORPHA. . (Genera—*Vampyrella, Archerina, Protomyxa,* etc.)

Phylum MYCETOZOA   . { 1. **SOROPHORA.**
2. **ENDOSPOREA.**
3. **EXOSPOREA.**

Phylum LOBOSA   .   . { 1. **NUDA.**
2. **TESTACEA.**

Phylum RETICULARIA   . { 1. **IMPERFORATA.**
2. **PERFORATA.**

Phylum LABYRINTHULARIA .   .   (Genera—*Labyrinthula, Chlamydomyxa*).

Phylum HELIOZOA   .   . { 1. **APHROTHORACA.**
2. **CHLAMYDOPHORA.**
3. **CHALAROTHORACA.**

Phylum
RADIOLARIA { Branch *a.* PORULOSA { 1. **PERIPYLARIA.**
2. **ACANTHARIA.**
Branch *b.* OSCULOSA { 1. **MONOPYLARIA.**
2. **PHŒODARIA.**

Sub-Grade *b.*—Corticata.

Phylum MASTIGOPHORA . { 1. **FLAGELLATA.**
2. **CHOANOFLAGELLATA.**
3. **DINOFLAGELLATA.**
4. **CYSTOFLAGELLATA.**

Phylum SPOROZOA   .   . { 1. **GREGARINIDEA.**
2. **COCCIDIIDEA.**
3. **MYXOSPORIDIA.**
4. **SARCOCYSTIDIA.**

Class.

Grade *a*. CILIATA.

Branch *a*. *Gymnostoma*.

1. **HOLOTRICHA.**

Phylum INFUSORIA

Branch *b*. *Trichostoma*.

1. **ASPIROTRICHA.**
2. **SPIROTRICHA.**

Grade *b*. SUCTORIA.

1. **ACINETARIA.**

Phyla and Classes of the *ENTEROZOA* (syn. *METAZOA*).

Sub-Grade *a*.—**Cœlentera.**

Class.

1. Phylum PORIFERA
  1. **CALCISPONGIÆ.**
  2. **SILICOSPONGIÆ.**

2. Phylum NEMATOPHORA.
  1. **HYDROMEDUSÆ.**
  2. **SCYPHOMEDUSÆ.**
  3. **ANTHOZOA.**
  4. **CTENOPHORA.**

Sub-Grade *b*.—**Cœlomata.**

Phylum PLATYHELMIA

Branch *a*. CILIATA
  1. **RHABDOCŒLA.**
  2. **DENDROCŒLA.**
  3. **NEMERTINA.**
  4. **ORTHONECTIDA.**
  5. **RHOMBOZOA.**

Branch *b*. COTYLOPHORA
  1. **TREMATOIDEA.**
  2. **CESTOIDEA.**
  3. **HIRUDINEA.**

Phylum ECHINODERMA

Branch *a*. AMBULACRATA
  1. **ECHINOIDEA.**
  2. **HOLOTHURIIDEA.**
  3. **ASTEROIDEA.**
  4. **OPHIUROIDEA.**

Branch *b*. TENTACULATA
  1. **CRINOIDEA.**
  2. **CYSTOIDEA.**
  3. **BLASTOIDEA.**

2 C

Class.

Phylum APPENDICULATA

Branch *a.* ROTIFERA
1. **PARAPODIATA.**
2. **LIPOPODA.**

Branch *b.* CHÆTOPODA
1. **POLYCHÆTA.**
2. **OLIGOCHÆTA.**
3. **MYZOSTOMARIA.**
4. **HAPLODRILI.**

Branch *c.* ARTHROPODA
Grade *a.* CERATOPHORA.
1. **PERIPATOIDEA.**
2. **MYRIAPODA.**
3. **HEXAPODA.**
Grade *b.* ACERATA.
1. **CRUSTACEA.**
2. **ARACHNIDA.**
3. **PANTOPODA.**
4. **TARDIGRADA.**

Phylum MOLLUSCA

Branch *a.* GLOSSOPHORA
1. **GASTROPODA.**
2. **SCAPHOPODA.**
3. **CEPHALOPODA.**

Branch *b.* LIPOCEPHALA
1. **LAMELLIBRANCHIA.**

Phylum GEPHYRÆA . .
1. **STERNASPIDOMORPHA.**
2. **ECHIUROMORPHA.**
3. **SIPUNCULOMORPHA.**
4. **PHORONIDOMORPHA.**
5. **POLYZOA.**
6. **PTEROBRANCHIA.**
7. **BRACHIOPODA.**

Phylum NEMATOIDEA .
1. **EUNEMATOIDEA.**
2. **CHÆTOSOMARIA.**
3. **CHÆTOGNATHA.**

Phylum GASTROTRICHA . Genera—*Chætonotus, Ichthydium.*
Phylum ACANTHOCEPHALA Genus—*Echinorhynchus.*

Class.

Phylum VERTEBRATA

Branch *a.* HEMICHORDA
: Sole Genus—*Balanoglossus.*

Branch *b.* UROCHORDA
: Grade *a.* LARVALIA.
: 1. **APPENDICULARIÆ.**
: Grade *b.* SACCATA.
: 1. **ASCIDIACEA.**
: 2. **THALIACEA.**

Branch *c.* CEPHALOCHORDA
: Sole Genus—*Amphioxus.*

Branch *d.* CRANIATA
: Grade *a.* CYCLOSTOMA.
: 1. **MYXINOIDEA.**
: 2. **PETROMYZONTIA.**
: Grade *b.* GNATHOSTOMA.
: Sub-Grade 1. BRANCHIATA HETERODACTYLA.
: 1. **PISCES.**
: 2. **DIPNOI.**
: Sub-Grade 2. BRANCHIATA PENTADACTYLA.
: 1. **AMPHIBIA.**
: Sub-Grade 3. PENTADACTYLA LIPOBRANCHIA.
: Branch *a. Monocondyla.*
: 1. **REPTILIA.**
: 2. **AVES.**
: Branch *b. Amphicondyla.*
: 1. **MAMMALIA.**

THE END

Printed by R. & R. CLARK, *Edinburgh.*